Fractals and Chaos

To Jim and Peta,
Best Wishes,
Tony
5. iii. 1991.

A.J. Crilly
R.A. Earnshaw
H. Jones
Editors

Fractals and Chaos

With 146 Figures in 173 Parts, 57 in Color

Springer-Verlag
New York Berlin Heidelberg London
Paris Tokyo Hong Kong Barcelona

A.J. Crilly
Middlesex Polytechnic
Barnet, Herts EN4 0PT
United Kingdom

R.A. Earnshaw
University of Leeds
Leeds LS2 9JT
United Kingdom

H. Jones
Middlesex Polytechnic
London N11 2NQ
United Kingdom

Cover illustration: Fractal planet based on the rescale-and-add method. For more details see Plate 25 in the colour insert (image by D. Saupe, University of Bremen).

Library of Congress Cataloging-in-Publication Data
Fractals and chaos / A.J. Crilly, R.A. Earnshaw, H. Jones, editors.
    p.   cm.
    1. Fractals. 2. Chaotic behaviour in systems.   I. Crilly, A.J.
II. Earnshaw, Rae A., 1944–   . III. Jones, Huw, 1944–   .
QA614.86.F7   1991
514'.74–dc20   90-37984

Printed on acid-free paper

©1991 by Springer-Verlag New York Inc.
All rights reserved. This work may not be translated or copied in whole or in part without the written permission of the publisher (Springer-Verlag New York, Inc., 175 Fifth Avenue, New York, NY 10010, USA), except for brief excerpts in connection with reviews or scholarly analysis. Use in connection with any form of information storage and retrieval, electronic adaptation, computer software, or by similar or dissimilar methodology now known or hereafter developed is forbidden.
The use of general descriptive names, trade names, trademarks, etc. in this publication, even if the former are not especially identified, is not to be taken as a sign that such names, as understood by the Trade Marks and Merchandise Marks Act, may accordingly be used freely by anyone.

Typeset by Nancy A. Rogers using TEX. Photocomposition on a Chelgraph IBX-2000.
Printed and bound by R.R. Donnelley & Sons, Harrisonburg, Virginia.
Printed in the United States of America.

9 8 7 6 5 4 3 2 1

ISBN 0-387-97362-1 Springer-Verlag New York Berlin Heidelberg
ISBN 3-540-97362-1 Springer-Verlag Berlin Heidelberg New York

Contents

About the Editors	vii
Preface	ix
Introduction	1

Part 1   Fractals

Chapter 1
Jones: Fractals Before Mandelbrot — 7

Chapter 2
Reeve: Mandelbrot, Julia Sets and Nonlinear Mappings — 35

Chapter 3
Batty: Cities as Fractals: Simulating Growth and Form — 43

Chapter 4
Kaandorp: Modelling Growth Forms of Sponges with Fractal Techniques — 71

Chapter 5
Saupe: Random Fractals in Image Synthesis — 89

Chapter 6
Horn: IFSs and the Interactive Design of Tiling Structures — 119

Chapter 7
Bressloff and Stark: Neural Networks, Learning Automata and Iterated Function Systems — 145

Part 2 Chaos

Chapter 8
Crilly: The Roots of Chaos—A Brief Guide 193

Chapter 9
Lansdown: Chaos, Design and Creativity 211

Chapter 10
Novak: Relativistic Particles in a Magnetic Field 225

Chapter 11
Mullin: Chaos in Physical Systems 237

Chapter 12
Darbyshire and Price: Phase Portraits from Chaotic Time Series 247

Chapter 13
Pottinger: Data Visualisation Techniques for Nonlinear Systems 259

Index 269

About the Editors
Fractals and Chaos

A.J. Crilly

Tony Crilly began his education in Sydney, Australia and later obtained undergraduate and masters degrees in mathematics from the University of Hull in the United Kingdom. He received his Ph.D. in the history of mathematics in 1981 from the Council for National Academic Awards. He has served as both secretary and treasurer of the British Society for the History of Mathematics and is currently a committee member. He is a Fellow of the Institute of Mathematics and its Applications and a member of the British Computer Society Documentation and Displays Group. He has taught in the United States at the University of Michigan and recently spent two years teaching at the newly established City Polytechnic of Hong Kong. While in Hong Kong, he helped set up a department of Applied Mathematics and designed courses in engineering mathematics and discrete mathematics. His present interests lie in the geometry of computer graphics and in mathematical education. He is currently Principal Lecturer in the Faculty of Engineering, Science and Mathematics at Middlesex Polytechnic. He is married with four children and lives in St. Albans, Hertfordshire.

R.A. Earnshaw

Rae Earnshaw is Head of Computer Graphics at the University of Leeds, with interests in graphics algorithms, integrated graphics and text, display technology, CAD/CAM and human-computer interface issues. He gained his Ph.D. in computer science at the University of Leeds. He has been a Visiting Professor at IIT, Chicago, USA, Northwestern Polytechnical University, China and George Washington University, Washington DC, USA. He was a Co-Chair of the BCS/ACM International Summer Institute on 'State of the Art in Computer Graphics' held in Sterling, Scotland in 1986; in Exeter, England in 1988; and in Edinburgh, Scotland in 1990. Dr. Earnshaw is also a Director of the NATO ASI on 'Theoretical Foundations of Computer Graphics and CAD' held in Italy in 1987.

H. Jones

Huw Jones was brought up in South Wales and graduated from University College Swansea with a B.S. in Applied Mathematics in 1966. The following year he obtained a Diploma in Education from the same institution and, after a short period as a schoolmaster, has spent the rest of his working life as a lecturer in higher education in London. During this period he obtained his Master of Science in Statistics from Brunel University, became a Fellow of the Royal Statistical Society, a member of the European Association for Computer Graphics and a member of the British Computer Society's Computer Graphics and Displays Group Committee. He is currently a Principal Lecturer specialising in Computer Graphics in the Faculty of Engineering, Science and Mathematics at Middlesex Polytechnic, where he is head of the Master of Science in Computer Graphics course. He is married to Judy, a mathematics teacher, and has a son, Rhodri, and a daughter, Ceri.

Preface

This volume is based upon the presentations made at an international conference in London on the subject of 'Fractals and Chaos'. The objective of the conference was to bring together some of the leading practitioners and exponents in the overlapping fields of fractal geometry and chaos theory, with a view to exploring some of the relationships between the two domains.

Based on this initial conference and subsequent exchanges between the editors and the authors, revised and updated papers were produced. These papers are contained in the present volume.

We thank all those who contributed to this effort by way of planning and organisation, and also all those who helped in the production of this volume. In particular, we wish to express our appreciation to Gerhard Rossbach, Computer Science Editor, Craig Van Dyck, Production Director, and Nancy A. Rogers, who did the typesetting.

A.J. Crilly
R.A. Earnshaw
H. Jones
1 March 1990

Introduction
Fractals and Chaos

The word 'fractal' was coined by Benoit Mandelbrot in the late 1970s, but objects now defined as fractal in form have been known to artists and mathematicians for centuries. Mandelbrot's definition—"a set whose Hausdorff dimension is not an integer"—is clear in mathematical terms. In addition, related concepts are those of self-similarity and sub-divisibility. A fractal object is self-similar in that subsections of the object are similar in some sense to the whole object. No matter how small a subdivision is taken, the subsection contains no less detail than the whole. Typical examples of fractal objects are 'Durer's Pentagons' (known to the artist Albrecht Durer in about 1500), the 'Pythagorean Tree' and the 'Snowflake Curve' (devised by Helge von Koch in 1904). The latter curve is a mathematical peculiarity which, if continually subdivided, produces a curve of infinite length which encloses a finite area. The distance along the curve between any two points is immeasurable—there is not enough wire in the world to bend into the shape of the Koch curve. These examples have exactly similar subsections, but many fractal objects, particularly those which occur naturally, have statistically similar subsections, so that subsections have similar forms with some variations.

Chaos is a topic that has developed through the study of dynamical systems and has connections with fractal geometry. Chaotic systems have the appearance of unpredictability but are actually determined by precise deterministic laws, just like many fractal images. Chaotic systems—often referred to as non-linear systems—show major fluctuations for apparently minor changes in the parameters which control them. The 'butterfly effect' is often used to illustrate the concept of a chaotic system. The breeze caused by the beating of a butterfly's wings may be the initial seed which eventually generates a hurricane. Similarly, minor changes in the parameters controlling the behaviour of a pendulum or the flow of a fluid could lead to major changes in the form of motion produced anywhere from smooth or laminar to chaotic. Minor changes in parameters thus cause major changes in the behaviour of chaotic systems. Such theories have been applied, for example, to meteorology, irregularities in heart beats, population modelling, quantum mechanics and astronomy.

This volume brings together a number of contributions in the areas of fractals, chaos and the interrelationship between the two domains. These contributions cover a wide variety of applications areas. This indicates the extent to which fractal and chaotic phenomena are being studied in the various disciplines. It is anticipated that this inter-disciplinary nature of the subject will increase, which in turn will yield useful information on the potential (and also limitations in some cases) of fractals and chaos as modelling tools for the investigation of various natural and scientific phenomena.

The volume is divided into two main sections. The first section contains contributions on fractals, the second material on chaos. In some cases the boundaries of the particular contribution are not clear cut, since contributions in both areas are being made. This indicates the interrelationship of fractals and chaos in these areas.

In Chapter 1, Huw Jones provides a selective history of fractals before Mandelbrot, which provides a general introduction and a definition of fractal dimension comprehensible to the layman. A number of examples are given to illustrate this definition.

This is followed in Chapter 2 by a survey of sets and nonlinear mapping by Dominic Reeve. He takes the standard iterative methods for producing fractals and applies them to functions other than those used to produce the classic Mandelbrot set. This could provide the basis for readers to develop and explore similar effects for themselves.

In Chapter 3, Michael Batty applies the ideas of fractal geometry to the spatial development of cities. Here, the processes of growth are initimately linked to the resulting geometry of the system. The model uses diffusion-limited aggregation to generate highly ramified tree-like clusters of particles or populations, with self-similarity about a fixed point. The model also takes into account constraints such as the limitations to the directions of growth imposed by natural phenomena, for example rivers, the sea and mountains. It is possible that these developments could have far-reaching implications for the future planning of cities and their populations. This in turn depends on the establishment of robust relationships between form and process.

The next contribution by Jaap Kaandorp in Chapter 4 outlines an interesting application in the area of sponges. This is based on an elegant, ramified fractal generation method which involves iteratively applying generators to an initial shape. Such generators can be randomly varied to produce different effects. Modelling sponges in this way raises the question of whether the model can be said to reflect actual biological processes and thereby give an insight into the actual mechanisms of physical growth.

In Chapter 5, Dietmar Saupe presents a number of methods for the generation of random fractals which have potential application in a wide variety of areas, e.g., visual images, sound, music, etc. The simulation of real world objects such as mountains, clouds, trees and plants is done in a remarkably convincing way in terms of our visual perception of these phenomena. A number of algorithms are presented which the readers may wish to use as a basis for experimentation. In addition, the extension of traditional space-oriented approaches to include the time dimension opens up possibilities for simulation and animation using these methods.

This is followed in Chapter 6 by Alastair Horn addressing tiling structures. This is based on the concept of Iterated Function Systems (IFSs) introduced by Hutchinson and later developed by Barnsley. The fractals resulting from such methods are deterministic, although random processes have been used as tools in their production. An interactive technique for the development of tiling structures is presented. This is implemented on parallel hardware and enables the points to be displayed almost simultaneously. The speed of processing and display allows real-time interaction by the user.

The final chapter in Part 1 (by Paul Bressloff and Jaroslav Stark) examines the underlying relationship between neural networks and IFSs. Possible applications to image generation, data compression and stochastic learning automata are discussed.

Part 2 of the volume is on Chaos. It is introduced in Chapter 8 by Tony Crilly, who gives a brief historical background to nonlinear systems and the evolution of ideas leading to an understanding of chaotic behaviour. Some elementary examples in the area of the physical sciences are used to illustrate the fundamental ideas of the modern approach to chaos theory. For those new to the field of chaos, the list of references provides useful background reading.

Complex systems often require a large number of variables and equations to satisfactorily represent and model them. One of the objectives of the study of chaos is to reduce these to a simpler set which still adequately models the system and whose behaviour can be computed in a much shorter time-frame.

In Chapter 9, John Lansdown presents a broad survey of the interplay between chaos and design, where the creative aspects of what is produced are important. From an artist's and designer's point of view, he explores those features of chaos which can be used to enhance the design process, often in unexpected directions.

This is followed in Chapter 10 by Miroslav Novak. He examines the application area of the motion of relativistic charged particles in a magnetic field, whose trajectories can be perturbed by travelling waves. The resulting motion can lead to chaotic dynamics, characterised by stable regions separated by stochastic boundaries. The images produced are striking in their aesthetic quality as well as their scientific value.

In Chapter 11, Tom Mullin investigates chaos in physical systems by looking at the areas of mechanics, electronics and hydrodynamics. Systems governed by deterministic equations can exhibit irregular or chaotic behaviour. The behaviour of a pendulum and a nonlinear electronic oscillator are used as the basis for examining the origins of turbulence in fluid flow. It is demonstrated that chaos does not arise because of noise in the system or imprecision of measurement. Instead it is an intrinsic feature of the physical system.

In Chapter 12, Alan Darbyshire and Tim Price show how an understanding of chaotic phenomena can be enhanced by illustration. Phase portraits are constructed from the time series resulting from dynamical systems which exhibit chaos. These are then presented graphically. Simple phase portraits are normally two-dimensional. Here, these are now extended to three dimensions. Future work is planned to extend this to even higher dimensions.

In the final Chapter, David Pottinger shows how data visualisation techniques can assist in improved understanding of nonlinear systems. Unpredictable effects in field theory can be modelled and viewed as three-dimensional animated sequences.

This volume has brought together a number of distinctive contributions in the areas of fractals and chaos. It is hoped that an understanding of fractals and chaos will lead to a common basis for examining the growth, development, organisation and behaviour of complex dynamical systems, many of which make up the natural world of which we are part. It is anticipated that the investigations of fractal structure associated with phase portraits will be an exciting area for future work.

While it is legitimate to consider fractals and chaos separately, the relationship between the two areas is important. The fractal nature of Julia sets is concerned with the 'chaotic' dynamics of iteration, while more realistic dynamical systems which display chaos can give rise to questions of both mathematical and philosophical significance and also to controversy. The importance of mathematical rigour and the role which it plays in the discovery of new ideas is once again being questioned. Chaos throws new light on the usefulness of mathematical models (as used in forecasting and prediction, for example). But the new methods are by no means universally accepted and are often fiercely debated in the academic world.

A.J. Crilly
R.A. Earnshaw
H. Jones
1 March 1990

1 Fractals

Fractals Before Mandelbrot
A Selective History

Huw Jones

Abstract

The word 'fractal' was coined by Benoit Mandelbrot in the late 1970s, but objects now considered as fractal in form existed or were invented before this date. This paper reviews some of those objects and describes their properties, including the concept of 'fractal dimension'. The intention is that the paper should serve as an introduction to the uninitiated through discussion of a variety of types of fractal objects. Those who have previous knowledge of fractals may find some forms previously unknown to them.

Introduction

The word fractal was coined by Benoit Mandelbrot in the 1970s. He dates the origin of 'Fractal Geometry' from 1975 [Mand88] but indicates that objects now considered as fractal existed long before that decade. Many naturally occurring objects, such as trees, coastlines or clouds, are now considered to have fractal properties; and much of the current interest in the topic stems from the attempts to simulate such natural phenomena using computer graphics. Other more abstract forms of fractal objects were devised by artists and mathematicians, and again the availability of current techniques of computer graphics has given new insight into the structure of such objects.

Mandelbrot's definition of a fractal set X is one whose "*Hausdorff dimension* $h(X)$ is not an integer"[Peit86]. We shall look more closely at this formal definition later but will be satisfied for now with a more intuitive idea of what comprises a fractal object. The essential property is one of 'self-similarity'. Subsets of a fractal object have essentially the same form as the whole. Theoretical fractal objects are infinitesimally subdivisible in this way, each subset, however 'small', containing no less detail than the complete set. Observations of fractal objects in reality are approximations to this ideal state, as at some stage of subdivision detail is inevitably lost. These concepts of self-similarity and infinitesimal subdivisibility are vague, but we shall proceed in the spirit of Karl Popper, who said "Never let yourself be goaded into taking seriously problems about words and their meanings ... any move to increase clarity or precision must be *ad hoc* or 'piecemeal' " [Popp76]. We shall look, in an ad hoc way, at several examples of fractal objects, mainly devised in the pre-Mandelbrot era, in order to gain insight as to their true structures.

Escher

Mandelbrot's article [Mand88] cites several historical examples of fractal objects, most developed by mathematicians in the early years of this century. An artist who is picked out for special attention is Dutchman Maurits Escher (1902-1972), whose tilings of plane figures seem intuitively to have some fractal form. It is interesting to read of the inspiration he received through correspondence with Henri Poincaré, who published widely on a variety of mathematical topics around the turn of the century. Douglas Hofstadter expresses the fractal nature of some of Escher's work in a section of his book [Hofs80] entitled *Copies and Sameness*: "Escher took the idea of an object's parts being copies of the object itself and made it into a print: his woodcut Fishes and Scales". Hofstadter's book, which links the works of Escher with the mathematical theories of Kurt Gödel and the music of Johann Sebastian Bach, also includes some fractal graphs which he calls INT and Gplot. In discussing Gödel's work, Hofstadter makes reference to an early publication by Mandelbrot, stating, "The boundary between the set of truths and the set of falsities is meant to suggest a randomly meandering coastline which, no matter how closely you examine it, always has finer levels of structure and is consequently impossible to describe exactly in any finite way." Thus, we are brought back full circle to Mandelbrot and to a convincing description of a fractal structure.

Dürer

A much earlier artist who generated a fractal object based on regular pentagons was Albrecht Dürer (1471-1528). If we take a regular pentagon of side length s and surround that by five identical pentagons, the shape created almost fits exactly into a larger pentagon with side length S [Dixo87]. The figure now resembles a large pentagon with five cutouts in the shape of isosceles triangles at the centre of each side (see Figure 1). The equal angles of these triangles are $72°$ (the external angle of a regular pentagon is $360°/5$), the third angle being half of this at $36°$ (as in triangle ABC of Figure 2) [Cund61]. The ratio of the sides of this triangle are in the 'golden section', well known since the days of Pythagoras (see [Cund61] and [Boye68] for further discussion). All that we need to know is that the smaller side of the triangle can be calculated as $2s \cos 72°$. This means that the ratio of the lengths of the sides of our two pentagons is

$$\frac{s}{S} = \frac{1}{2 + 2\cos 72°}$$

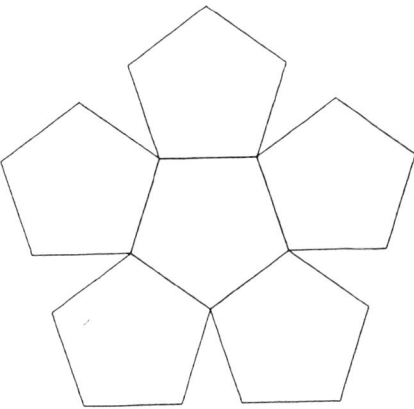

Figure 1. The basis of Dürer's pentagons.

Now imagine starting the procedure from the large pentagon. Knowing the way to calculate s from S, we could draw the six smaller pentagons which sit inside the large one. Why be satisfied with this? The subdivision can continue, with ever decreasing pentagons being produced (Figure 3). Each subpentagon is a copy of the whole. If subdivision is continued ad infinitum, Dürer's pentagons form a truly fractal object.

Cantor

During the late nineteenth century the theory of sets was being developed. Mathematicians delighted in producing sets with ever more weird properties, many of them now recognised to be fractal in nature. One of these is the set devised by Georg Cantor (1845–1918) [Lauw87]. Its construction is relatively simple and can be illustrated by the Cantor comb (Figure 4). Begin with all real numbers in the interval [0, 1] of the real line. Extract the interval (1/3, 2/3) which constitutes the central third of the original interval, leaving the two closed intervals [0, 1/3]

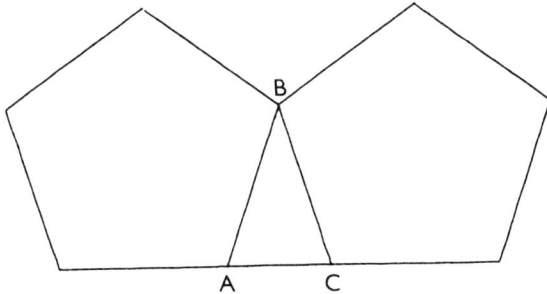

Figure 2. The golden triangle ABC cut out of Dürer's pentagons.

Figure 3. A Dürer pentagon after four subdivisions.

and [$2/3, 1$]. Continue this process, at each stage extracting the central third of any interval that remains. If this is continued ad infinitum, the remaining points rest on the edges of the teeth of the 'comb' illustrated. This may not seem particularly remarkable at first, but the set has some unusual properties. The infinite series corresponding to the lengths of the extracted sections forms a simple geometric progression

$$\left[1 + \left(\frac{2}{3}\right) + \left(\frac{2}{3}\right)^2 + \left(\frac{2}{3}\right)^3 + \cdots\right]\bigg/ 3$$

as at stage one, an interval of length $1/3$ is cut out; on remaining stages twice as many intervals, each $1/3$ the length of the previous cut out intervals, are extracted. A piece of relatively simple high school arithmetic shows that this sums to unity, meaning that the points remaining in the Cantor set, although infinite in number, are crammed into a total length of magnitude zero. Such points must be disconnected; there is some unfilled space between any pair of points in the set, no matter how close these points may be. The set is said to form a 'dust'.

Numbers in general can be represented with respect to any number base. As the construction of the Cantor set involves repeated division by 3, it is informative to represent these numbers to the base 3. A typical value in the interval [0, 1] could be represented in this form as a sequence of the digits 0, 1 and 2, for example

$$0.201120210_{(\text{base }3)}$$

The value illustrated here could not belong to the Cantor set. As the central third of each existing interval is discarded, the digit 1 cannot occur in the base 3 representation of a value in the Cantor set. Only the digits 0 and 2 can occur, for example

$$0.202220220_{(\text{base }3)}$$

The digits 0 and 2 can be used to trace a path down the Cantor comb to the required value, 2 indicating that the right 'tooth' is taken, 0 indicating the left

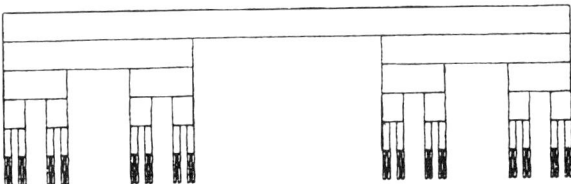

Figure 4. A Cantor comb after five subdivisions.

'tooth'. As values in the Cantor set can be represented in this form, there is a one-to-one correspondence between numbers in the Cantor set and the binary numbers in the range [0, 1]. For instance, changing all the 2's to 1's in our example above gives us a binary number

$$0.101110110_{(\text{base 2})}$$

Such binary numbers, familiar as the method used to store numbers in a digital computer, represent all real values in the interval [0, 1]; thus they 'fill' this interval. Thus, the Cantor set contains an infinite set of values which lie in a zero length space, yet they have a one-to-one correspondence with the set of all real values which fill the interval [0, 1].

Much of Cantor's work was involved with unravelling the paradoxes of the concept of infinity. One of his contemporaries, J.W.R. Dedekind (1831–1916), stated a universal property of infinite sets which sounds very much like our concept of fractal objects today: "A system S is said to be infinite when it is similar to a proper part of itself." [Boye68].

Peano, Hilbert and von Koch

In 1890 Giuseppe Peano (1858–1932) "showed how thoroughly mathematics could outrage common sense when he constructed continuous space filling curves" [Boye68]. David Hilbert (1862–1943) later developed a similar construction, a curve which visits every point in a square and which is nowhere differentiable, leading Carl Boyer to comment, "it was in the twentieth century that anomalies and paradoxes ran rampant" [Boye68] (see also Dietmar Saupe [Saup88] for an illustration of this curve). The curve generated by Helge von Koch (1870–1924) in 1904 is typical of the type of curve generated at that time and is one of the classical fractal objects (Figure 5). The curve is constructed from a line segment, which we can take, without loss of generality, to be of unit length. Then the central third of this line is extracted and replaced by two lines of length $1/3$, as in Figure 5. This process is continued, the central third of any line segment being replaced at each stage by two lines of length one third that of the segment. The protrusion of the replacement is always on the same side of the curve. Note that the points at which the final curve touches the original line are the points of the Cantor set. Consider the effect of this process as the number of stages

Figure 5. Four stages in the generation of a Koch curve.

increases. At each stage, the total length of the curve is multiplied by $4/3$. A little exercise with a calculator should be enough to convince any skeptic that repeated multiplication of unity by $4/3$ creates a number too large to store in the calculator; in fact the length of the completed Koch curve is infinite. Now consider the area between the Koch curve and the original line. At the first stage one triangle is added, its area being equal to

$$A = \left(\frac{1}{2}\right)\left(\frac{1}{3}\right)\left(\frac{\sqrt{3}}{6}\right) = \frac{\sqrt{3}}{36}$$

using the familiar formula 'half times base times height' for the area of a triangle. The actual value of A is immaterial for this development; it is given here merely to demonstrate how easily it can be calculated. At the next stage four new triangles are added, their sizes being linearly scaled down by a factor of $1/3$. Thus their areas are $(1/3)^2 A = (1/9)A$. At each stage, four times as many new triangles are added to the total area, the area of each triangle added being one-ninth those added at the previous stage. Thus, the increase in area at any stage is four-ninths the area added at the previous stage. This leads to a geometric progression for the total area

$$A\left(1 + \frac{4}{9} + \left(\frac{4}{9}\right)^2 + \left(\frac{4}{9}\right)^3 + \cdots\right)$$

which has the finite sum

$$\frac{9A}{5} = \frac{\sqrt{3}}{20}$$

Thus, we have a curve of infinite length which encloses a finite area. This curve, like Peano's and Hilbert's, is nowhere differentiable, that is, it does not have a well defined slope at any point. Also, it contains an infinite number of perfect miniature images of itself. As the curve is infinite in length, any scaled down subimage is also of infinite length. This leads to the conclusion that for any two points on the curve, no matter how 'close' they are, the curve between them is of infinite length.

Another curve with interesting properties can be constructed as a sequence of paper folds. Take a long strip of paper and fold it in half. Continue folding the remaining sections in half, always folding in the same direction. When you can fold no further, unravel the paper strip, making each corner a right angle in the natural direction of the fold. The resulting shape is the dragon curve (Figure 6). The curve gives a tight spatial fit but never crosses itself. Its name is probably due to the resemblance of its exterior boundary to traditional Chinese 'dragons'. If the corners are numbered 1, 2, 3, ... from one end of the curve, the rule for the direction of the bend for corner i can be given as follows:

if i is even, turn in the same direction as for corner $i/2$;
else if the remainder on dividing i by 4 is 1, turn right;
else if the remainder on dividing i by 4 is 3, turn left.

There are many other fractal curves which have similarly intriguing properties, for example, Levy's curve, Koch's quadric curve, the monkey tree curve, etc. For descriptions of these and others, see Lauwerier and Kaandorp [Lauw87] and Saupe [Saup88].

Sierpinski

Waclaw Sierpinski (b 1882) gave his name to a number of fractal objects, the Sierpinski arrowhead (or triangle or gasket) and the Sierpinski carpet, which are based in two-dimensional space, and the Sierpinski tetrahedron and sponge,

Figure 6. A dragon curve after 15 folds.

based in three-dimensional space. To construct a Sierpinski triangle, extract from an original triangle the inverted half-scale copy of itself formed by joining the midpoints of the three sides. Three half-scale triangles now remain, so one-fourth of the area of the original triangle has been removed. The process is now repeated for each triangle remaining in the object. At the second stage, one-fourth of the area of three triangles is removed, each of which is one-fourth of the area of the original; at stage three there are nine triangles removed, of area $(1/4)^3$ of the original triangle's area. If that original area is set to A, the area removed by this process gives another geometric progression

$$A\left[\frac{1}{4} + 3\left(\frac{1}{4}\right)^2 + 3^2\left(\frac{1}{4}\right)^3 + 3^3\left(\frac{1}{4}\right)^4 + \cdots\right] =$$

$$A\left[1 + \frac{3}{4} + \left(\frac{3}{4}\right)^2 + \left(\frac{3}{4}\right)^3 + \cdots\right]\bigg/4$$

which sums to A. As in the Cantor set, we have extracted a region of the same size as the whole of the original space, but we still have points left in the Sierpinski triangle. These points which exist in an area of magnitude zero are separate, forming a dust. In practice, the set can only be drawn to a given number of subdivisions, as in Figure 7.

A similar procedure can be performed using a square as the original region, subtracting a square at one-third scale from the centre and leaving eight sub-squares behind. The central one-third scale square of each of these can then be subtracted in a process that will, by now, be familiar. This gives a 'Sierpinski carpet', as in Figure 8.

The Sierpinski tetrahedron is created by a similar repetitive contraction and replacement strategy. The starting figure is a tetrahedron (or triangular based pyramid), which is replaced by four tetrahedra at half-scale, in the same orientation as the original tetrahedron and each having one vertex coincident with a vertex of the original. The four new tetrahedra lie completely within the space previously occupied by the original tetrahedron, the portion eliminated forming the shape of an octahedron (a polyhedron with eight triangular faces). If the starting tetrahedron is regular, then the octahedron subtracted will be regular. This process, when repeated, reduces the tetrahedron to a dust of disconnected points by discarding half the existing volume at each stage (see Figure 9).

The Sierpinski sponge, also known as the Menger sponge, is a close relative of this shape. Its formation is very similar to that of the Sierpinski carpet, except that the starting shape is a cube in three dimensions rather than a square in two dimensions. Each square face of the cube is treated in exactly the same way as the original square of the Sierpinski carpet. This time, extracting a square shape involves punching a hole directly through the cube at right angles to the face concerned. Thus, at the first stage three holes are punched through (imagine taking away the centre subcube of a Rubik's cube together with the central

Figure 7. A Sierpinski triangle or gasket after five subdivisions.

subcube of each face). This leaves 20 subcubes at one-third scale, each of which is repeatedly subdivided to create the final fractal object.

Gaskets and Barnsley's Chaos Game

Michael Barnsley has shown how to create the Sierpinski gasket or triangle using his 'chaos game' [Barn88a, Barn88b]. The vertices of the triangle are taken as 'attractors'. Taking any starting point, one of the vertices is chosen at random, and a point is plotted half way between the starting point and the chosen vertex. This point is then taken as the starting point, a new point being plotted half way

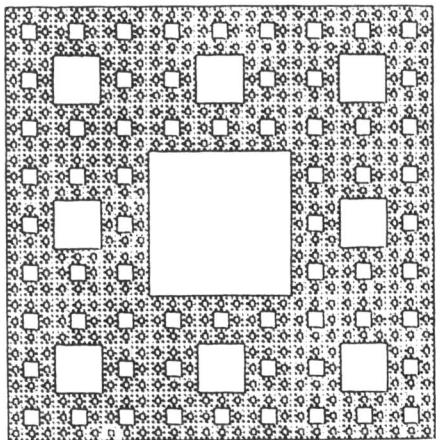

Figure 8. A Sierpinski carpet after five subdivisions.

16   Huw Jones

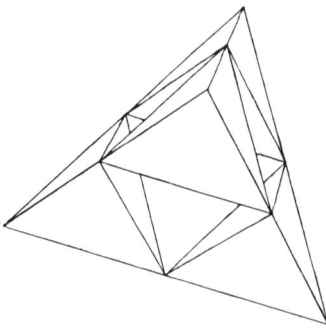 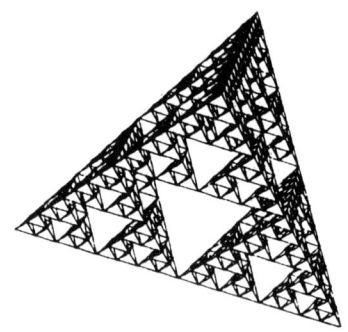

Figure 9. Two stages in the creation of a Sierpinski tetrahedron.

between it and a randomly chosen vertex. When several thousand points have been plotted—more easily done on a computer than using pencil and paper, even on a rainy day—the familiar shape of the Sierpinski gasket is found. Figure 10 shows the result of plotting 20,000 points in this way, using one of the triangle's vertices as the starting point. By starting at a vertex which lies in the gasket, we are guaranteed that all remaining points plotted lie within the required gasket.

By changing the attraction rule and starting with different patterns of points, many different types of 'gasket' can be produced. Suppose we start with vertices at the corners of any regular polygon with n sides. Imagine that the polygon has very small copies of itself placed within it at each of its vertices. Now make the small copies grow, still placed at the vertices and still remaining equal to each other, until they just touch. If you consider this to be done with a pentagon, the shape produced will be the generating shape for the Dürer pentagons (Figure 1). To produce a fractal gasket, the contents of the original polygon need to be shrunk to fit inside the subpolygons; thus the size of these copies when compared with the size of the original polygon is important in defining the 'attraction rule'. In evaluating this size, the exterior angle of the polygon, $360°/n$, is needed to decide the length of the portion to be cut out of the side of the original polygon. If $360°/n$ is not less than $90°$, that is, n is 4 or less, the edges of the smaller polygons meet exactly at the edge of the original polygon (triangle or square in

Figure 10. A Sierpinski gasket created using Barnsley's 'chaos game'.

these cases), as in Figures 11 and 12. When n is 5 or more, the smaller copies touch at a point inside the original polygon (as in Figure 1). When n is 5, 6, 7 or 8, the shape cut out at an edge is a triangle; for larger values of n, more complicated shapes are cut out. The determining value is

$$k = \mathrm{trunc}\left(\frac{90}{360/(n-1)}\right) = \mathrm{trunc}\left(\frac{(n-1)}{4}\right)$$

where $\mathrm{trunc}(x)$ is the truncated integer part of the real value x. The cut out length is given by

$$2s\sum_{i=1}^{k}\cos\left(\frac{360i}{n}\right) = 2sc$$

The value of k determines how many sides of the subpolygon are projected onto the cut out part of the original side (Figure 13 illustrates the situation for a nonagon cutout). Thus, if S is the side length of the original polygon we have

$$S = 2s + 2sc = 2s(1+c)$$

The required shrink factor is

$$f = \frac{s}{S} = \frac{1}{2(1+c)}$$

The gasket can then be constructed from an original point P by selecting one of the vertices V of the original polygon at random and then plotting a new point P' at a fraction f of the distance from V to P. If P is any point in the original polygon, this places P' in the scaled down copy which lies next to vertex V. For the triangular gasket, f will be $1/2$, as specified in Barnsley's method. Calculation of the position of P' is simple. Merely apply the formula

$$P' = (1-f)V + fP$$

to the x, y and z coordinates of V and P in turn. Some results of this process are shown in Figures 14 to 16. Note that the pentagonal based process (Figure 14)

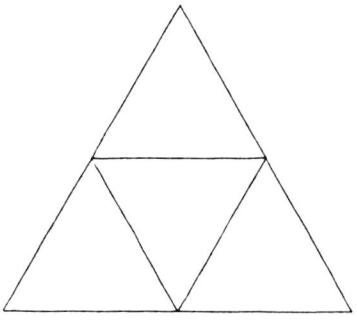

Figure 11. Subdivision of a triangle into half-scale subtriangles.

Figure 12. Subdivision of a square into half-scale subsquares.

gives a result similar to the Dürer pentagons, with the central pentagon left blank at each stage (Figure 17). The hexagonal form (Figure 15) has a number of internal and external bounding lines which take a familiar form—the Koch curve. Other gaskets are bounded by similar fractal curves. For large numbers of sides in the original polygon, the pattern produced becomes narrower, resembling a laurel wreath (Figure 16). If this procedure is attempted with a square, the resulting figure has a set of random points evenly distributed within the square, as the four half-size subsquares completely fill the original square. Setting f to a value less than $1/2$, for example $f = 1/3$, gives a shape like a 'nonfilled-in' Sierpinski carpet (Figure 18).

Pythagoras Tree

Most school text books on mathematics contain an illustration rather like a square tree trunk with two abbreviated branches to illustrate the proof of Pythagoras' theorem. Although the theorem is attributed to Pythagoras and was possibly justified by him, the proof based on this figure appears in Book 1 of *The Elements* by Euclid of Alexandria. The outline of the figure is used to create a fractal structure now called Pythagoras tree [Lauw87]. Copies of the outline are scaled, rotated and translated to create branches on the original trunk, creating after a number of iterations tree-like structures similar to the ramifying fractals

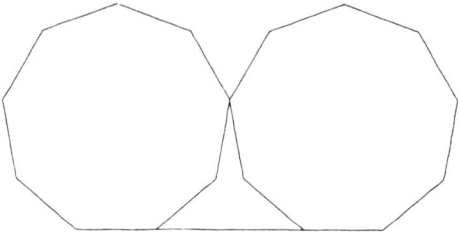

Figure 13. A nonagon 'cutout' of pentagonal form.

Figure 14. A pentagonal gasket.

of Jaap Kaandorp [Kaan90] (Figure 19). Whilst the common name of this fractal structure suggests a long pre-history, the suspicion is that the form is relatively recent, the name arising only from similarity with the schoolbook diagram that is used in the proof of Pythagoras' theorem.

Julia and Fatou

Fractals generated from the theories of Gaston Julia (1893—1978) and Pierre Fatou (1878—1929), dating from about 1918, are based in the complex plane. For the uninitiated, Courant and Robbins [Cour69] give a most readable introduction to the algebra of complex numbers, as, indeed, they do to most of the fundamental principles of mathematics. Their book, *What is Mathematics?* reviews much of the mathematical background required for computer graphics, although its first publication predates the development of that subject by some twenty years. It also contains many examples of fractal objects, including an intriguing

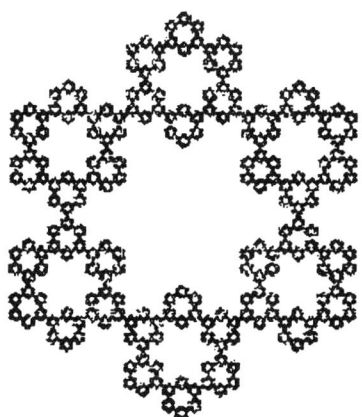

Figure 15. A hexagonal gasket.

20 Huw Jones

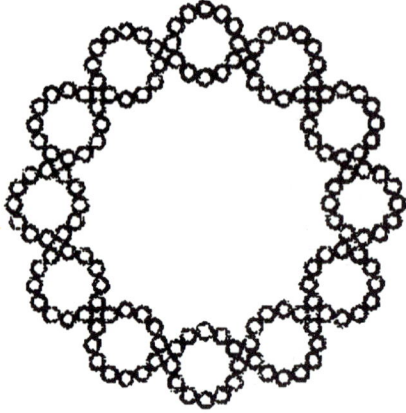

Figure 16. A dodecagonal gasket.

illustration of reflections in three cylindrical mirrors, although, of course, the word 'fractal' was not used to describe them.

Work similar to that of Julia and Fatou had been attempted earlier but with little success. For example, Arthur Cayley (1821—1895) was defeated by the complexity of attempting to determine which root of a complex equation would be approached from various starting points, using Newton's iterative method. Using modern computer techniques, it is relatively easy to demonstrate that the boundaries between the regions defined in this problem are fractal in nature. But without the concept of fractals and without computer power this problem proved too complicated for Cayley in 1879 [Peit86]. It was eventually solved by John Hubbard 100 years after Cayley's attempt [Glei88].

Complex numbers take the form $z = x + iy$, where the imaginary number i is defined such that
$$i^2 = -1$$

Figure 17. Dürer's pentagons with blank central section.

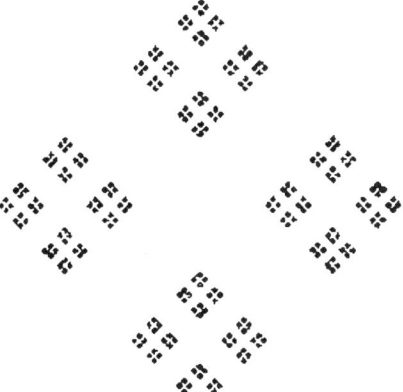

Figure 18. Chaos game based on a square with shrink factor of $1/3$.

The z values can be represented as points (x, y) in a plane, where the real value x represents the horizontal displacement from some fixed point called the origin, and the real value y represents the vertical displacement from that point. An algebra of complex numbers exists, with laws for addition, multiplication, etc. For example, using $i^2 = -1$, we can define

$$z^2 = (x + iy)(x + iy) = (x^2 - y^2) + i(2xy)$$

Suppose $c = a + ib$ is a complex constant. We then have

$$z^2 + c = (x^2 - y^2 + a) + i(2xy + b)$$

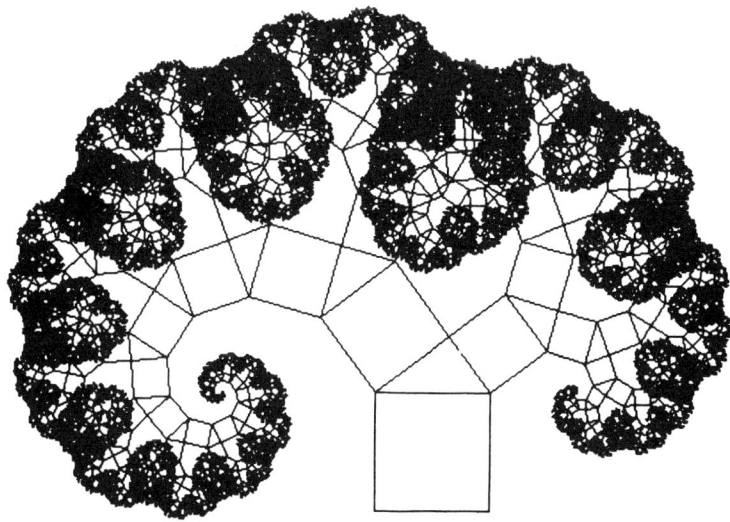

Figure 19. A Pythagoras tree based on a 3, 4, 5 triangle.

If we start with a pair of values (x_0, y_0) determining a point in the plane which is a representation of $z_0 = x_0 + iy_0$, we can use the function

$$z_1 = f(z_0) = z_0^2 + c$$

to create an 'image point' of (x_0, y_0) at

$$(x_1, y_1) = (\{x_0^2 - y_0^2 + a\}, \{2x_0y_0 + b\})$$

We have already seen that the creation of a fractal object often involves repetition of a particular rule, amending the starting point at each stage. This case is no exception. We now reapply the creation rule to (x_1, y_1) to find a second image point (x_2, y_2). This process is repeated. There are two frequently occurring outcomes of this process. Either the sequence of image points diverges from the origin, that is, the distances of the image points from the point $(0, 0)$ become increasingly large, eventually approaching infinity; or the sequence of points converges to a fixed point, getting closer and closer to that point. There are other possibilities, such as a finite cycle of points being repeated, but the divergent and convergent sequences are the most frequently observed. If c is fixed, points (x_0, y_0) which act as starting points for such an operation can be divided into two sets, the set of points for which the sequence diverges, and the set for which the sequence does not diverge. Those points which lead to nondivergent sequences but which lie next to points which create divergent sequences constitute the Julia set for a particular value of c. The Julia set is the boundary between the nondiverging values of (x_0, y_0) and the diverging values of (x_0, y_0), which lie just within the set of nondiverging values. It can be thought of as 'the thin red line' defending the nondiverging set from the diverging set. The complete set of nondiverging starting values of (x_0, y_0) is called the 'filled-in' Julia set [Doua86].

This set is fractal in nature, illustrating the difficulty that its early explorers had in defining it. The wonder of the development of Julia and Fatou is that they had none of the modern tools of computer graphics to illustrate their work. They developed abstract constructs which they never saw illustrated. It needed the insight of Mandelbrot to see the way in which their work could create fascinating images.

The direct method for computer graphic generation of a Julia set on a raster device involves repeated evaluation of the function $f(z) = z^2 + c$. The region of the complex plane for which the illustration is required is decided upon, enabling the (x_0, y_0) coordinates corresponding to the centre of each pixel of the device to be evaluated. For each pixel, using its (x_0, y_0) 'address' to generate

$$z_0 = x_0 + iy_0$$

as a starting value,

$$z_1 = f(z_0)$$

is calculated. If the distance of z_1 from the origin is larger than some arbitrarily 'large' value (10 is recommended by Peitgen and Richter [Peit86]), the decision is that the point (x_0, y_0) lies outside the Julia set (it is more sensible to use the square of this distance, $x_1^2 + y_1^2$, for this comparison against the square of the 'large' test value, thus saving computing time through nonevaluation of the square root). Otherwise, the calculation of

$$z_n = f(z_{n-1})$$

is repetitively performed, each time using the result of the previous stage's calculation as input into the function $f(z)$. After a set number of repetitions of this procedure (200 is recommended by Peitgen and Richter [Peit86]), if the distance of $f(z_n)$ from the origin is still not large, the conclusion is that the point (x_0, y_0) corresponding to the pixel lies within the 'filled-in' Julia set, and its pixel is accordingly allocated a colour. Points for which the value of $f(z_n)$ eventually does become 'large' according to our criterion are outside the Julia set. Their corresponding pixels can either be given a constant colour code or can be coloured according to the number of stages n it took to reach the 'escape value', which produces some fascinating patterns.

This method is computationally intensive and notoriously slow for high precision images requiring on the order of one million such repetitive calculations. A more subtle approach depends upon consideration of what happens to starting points which lie close to the required Julia set. If we start with a point (x_0, y_0) which lies just outside the Julia set, the sequence of points generated diverges away from the set. If we start with a point just inside the Julia set, the sequence of points always remains within the set, usually converging towards a fixed point somewhere inside the filled-in Julia set. If we could reverse this process we would create a sequence of points which would approach the boundary of the Julia set, regardless of our starting point. If the starting point (x_n, y_n) representing z_n were outside, the sequence $z_{n-1}, z_{n-2}, \ldots$ would approach the outer boundary; otherwise the sequence would approach the inner boundary. The problem is to find a way of reversing the sequence of points. Our original sequence depended on the evaluation of z_n from z_{n-1} using

$$z_n = z_{n-1}^2 + c$$

To reverse the procedure we need to find the value of z_{n-1} for a given value of z_n. The two possible values are

$$z_{n-1} = +\sqrt{z_n - c} \quad \text{or} \quad -\sqrt{z_n - c}$$

The method for finding the Julia set proceeds as follows: Given an arbitrary starting point z_n, calculate a value of z_{n-1}, selecting the plus or minus sign at random. Repeat this procedure a set number of times, at each stage using the calculated value from the previous stage as the starting point for the calculation. The final value of z_0 calculated should represent a point close to the boundary of the Julia set. A pixel corresponding to that value of z_0 can be coloured accordingly. The whole process is repeated a large number of times (typically several thousand), the random selection of a plus or minus sign at each stage of the process leading to many different points being plotted. If enough points are coloured in this way, the required Julia set begins to take shape. Frequently, this outline takes the form of the rugged coastline of an island (Figure 20). Other forms resemble fantastic fireworks displays in the sky (see Figure 21 and also [Peit86] and [Reev90]).

Different values of the constant c give many different shapes of Julia set which fall into two essentially different forms, connected and not connected. If a Julia set is connected, then any point within the set can be joined to any other point within the set by a line (possibly curved) which lies completely within the set—this is the coastline type described above, with the boundary enclosing a single 'island'. Other values of c give disconnected sets, reducing to a 'dust' of separate points in two-dimensional space, reminiscent of the Sierpinski triangle and the Cantor set [Peit86; Doua86; Reev90].

Julia and Fatou developed the theory of such sets in 1918 [Doua86], long before computer graphic techniques were available to display the images that fascinate us today. They developed abstract theories of these constructions that they never saw illustrated. Mandelbrot's inspiration was in using computer graphics to illustrate such sets, leading to computer generation of the now famous set which bears his name. The basis of this work is the complex function

$$f(z) = z^2 + c$$

There is no reason why such techniques should not give equally interesting results with other functions. Dominic Reeve [Reev90] uses both of the methods suggested above and shows how changing the power of z gives different effects. When using the inverse method for other powers of z there is a greater choice of values for continuing the sequence. For example, using the function

$$f(z) = z^k + c$$

gives k possible values of z_{n-1} for each z_n, so the random selection at each stage is from a larger set of possibilities.

Clifford Pickover has experimented with a wide range of functions, producing many fascinating images resembling primitive biological life forms which he calls biomorphs [Pick86].

Mandelbrot

Although this paper is intended as a 'pre-history' of fractals, it is impossible to end without at least a brief description of the set which bears Benoit Mandelbrot's name. This has been very well documented and illustrated elsewhere (for example, in Peitgen and Richter [Peit86]), so the discussion here will be brief.

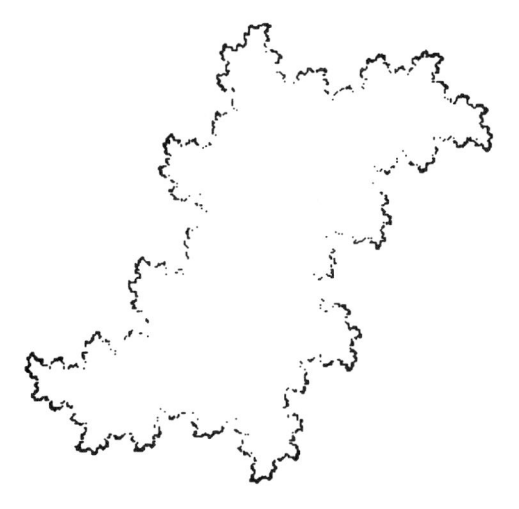

Figure 20. A connected Julia set for $c = 0.1 - 0.6i$, using 20,000 randomly selected points.

The Mandelbrot set can be defined from its relationship with Julia sets. The preceding section describes two forms of Julia set depending on the value of the constant c chosen: connected, and disconnected. The Mandelbrot set is the set of values of c for which the Julia sets are connected. If a point within the now familiar crinkly outline of the Mandelbrot set is used to define the constant c for generation of a Julia set, then the Julia set will be connected [Peit86; Reev90].

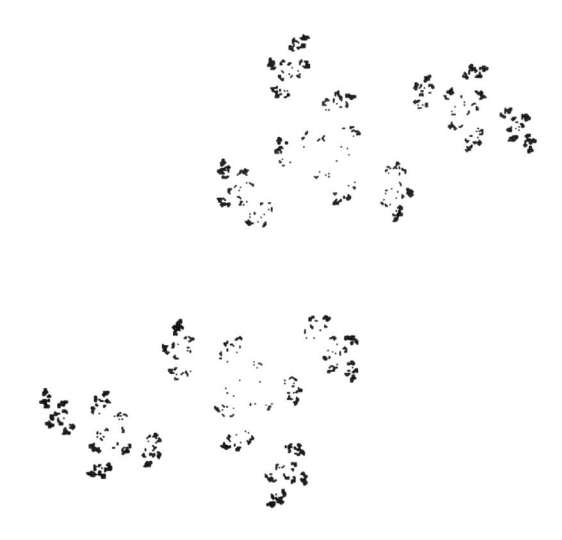

Figure 21. A disconnected Julia set for $c = 0.1 - 0.8i$, using 20,000 randomly selected points.

There is an alternative and equivalent definition which is usually used to create images of the Mandelbrot set. Adrien Douady explains that this equivalence is the result of a theorem proved independently by Fatou and Julia in 1919 [Doua86]. The resulting method is similar to that used for illustrating Julia sets on a raster device, when the function

$$f(z) = z^2 + c$$

is used to create sequences of complex values $z_0, z_1, z_2, \ldots$ corresponding to points in the (x, y) plane for a given value of c, the pixel corresponding to the starting point z_0 being coloured according to the properties of the particular sequence. To create an image of the Mandelbrot set, the starting point z_0 is set to zero, corresponding to the origin of the complex plane $(0, 0)$. Sequences $z_0, z_1, z_2, \ldots$ are evaluated in the same way as for the Julia set, except that the pixel position is made to correspond to the constant c. A region of the complex plane is chosen to correspond to the section of the raster display to be used, and a value of c is calculated corresponding to the centre of each pixel. Then, if the sequence defined above converges, the point corresponding to c lies within the Mandelbrot set. If it diverges, the point corresponding to c is outside the Mandelbrot set, and its pixel can be colour coded according to the number of stages the sequence took to reach the 'escape distance' from the origin.

Images of various regions of the complex plane are now familiar [Peit80], although it was only in March 1980 that Benoit Mandelbrot obtained the first hazy images of the Mandelbrot set on a fairly primitive output device [Mand86]. Some regions have become popular favourites, for example 'seahorse valley'. The familiar fractal property of infinitesimal subdivisibility is a feature of these images. If regions are subdivided into smaller and smaller parts, there is no loss of detail; more and more intricate patterns are revealed. At first glance the set appears to have a number of outlying islands whose structure is similar to the main region. However, it has been shown that the Mandelbrot set is connected; these outlying regions are joined to the main part by fine tendrils which are revealed at higher precision. This contrasts with the way in which an art historian may study the brush strokes of a master painter. Using a magnifying glass on a Seurat reveals the pointillist technique in detail. But if the power of the magnification is increased, a stage is eventually reached where there is no more detail to be revealed; the microscope focuses on a region of one single colour. This cannot happen with images of the Mandelbrot set and other fractal constructs, as detail continues to an infinitesimal scale.

Fractal Dimension

We now return to Mandelbrot's definition of fractals quoted in the first section of this paper, in the hope that exposure to fractal objects will lead to better understanding of the concept of fractal dimension. In the Euclidean or Newtonian universe dimensionality is clearly defined. A point has dimension zero; a line, whether straight or curved, has dimension one. Any point on the line can be represented by a single parameter, for example its distance along the line from a fixed point. Similarly, points on a plane or any normal curved surface can be represented by two values, for example (x, y) coordinates in a Cartesian plane, or latitude and longitude on the surface of a globe, making such surfaces two-dimensional. The Newtonian description of the space we live in is three-dimensional. Any point in our universe can be located by three values, for example the Cartesian coordinates (x, y, z). Mathematicians have constructed other spaces by analogously extending the laws governing lower dimensional spaces into higher dimensions (Abbott's book [Abbo87] contains an amusing discussion of such ideas). Higher dimensional spaces cannot be visualised, but they have proved useful in a number of contexts, for example in the development of rational splines [Pieg87]. The dimensions of all such spaces are nonnegative integers, 0, 1, 2, 3, ...

Some of the more complicated sets we have discussed above do not sit naturally in such restricted spaces. Consider the Cantor set. It consists of a set of disjoint individual values so could be thought of as having dimension zero. Yet it has a one-to-one correspondence with the set of all values on the real line in the interval [0, 1], which is fairly evidently a one-dimensional set. The Koch curve may appear to be one-dimensional according to the argument above. Yet points along the curve cannot be labelled in the normal way by distances from a fixed point, as the distance between any two points on the Koch curve is infinite. The Koch curve does not fill the two-dimensional space in which it exists, so it cannot be two-dimensional. Yet, it seems to be more than one-dimensional. Similar problems occur when considering the Sierpinski gasket and fractal objects which exist in three-dimensional space, such as the Sierpinski sponge and tetrahedron. Courant and Robbins [Cour69] discuss the problems that Poincaré had with this concept, but they do not suggest that anything other than a whole numbered dimension could exist.

There seems to be a need for a fractional dimension to describe the structure of such constructions; this is provided by the 'fractal dimension', or the 'Hausdorff dimension' [Voss88]. In order to approach this, consider the way in which dimension can be described for some more normal figures, for example a square or a triangle. These exist in two-dimensional space and are properly two-dimensional. We shall be discussing this in the context of fractal objects, which have the property of self-similarity—that is, small sections of the objects contain, in some sense, scaled down versions of the whole object (consider the analogous way in which whole plants can be cloned from small scraps of leaf material). Squares and triangles can be subdivided in this way to contain scaled down copies of themselves (Figures 11 and 12). Both squares and triangles can be subdivided into four copies at $1/2$ scale, nine copies at $1/3$ scale, 16 copies at $1/4$ scale. In each case, if the number of copies is N and the scale factor is f we have the relationship

$$N = \left(\frac{1}{f}\right)^2$$

A cube can be subdivided into eight copies at $1/2$ scale, 27 copies at $1/3$ scale (consider Rubik's cube), ... Now the relationship between N and f becomes

$$N = \left(\frac{1}{f}\right)^3$$

The power of $1/f$ indicates the dimension of the object. We can generalise. If an object can be subdivided into N copies of itself at scale f, then its dimension is the value D which satisfies

$$N = \left(\frac{1}{f}\right)^D$$

D can be obtained directly by taking logarithms as

$$D = \frac{\log(N)}{\log(1/f)}$$

Consider this when applied to some of the fractal objects generated earlier. The Cantor set contains two copies of itself at $1/3$ scale, thus its dimension is

$$D = \frac{\log(2)}{\log(3)} = 0.6309$$

to four decimal places, a noninteger dimension between 0 and 1, which is what our reasoning above would lead us to expect. For the Koch curve we have four copies at $1/3$ scale. Thus

$$D = \frac{\log(4)}{\log(3)} = 1.2619$$

to four decimal places, a noninteger dimension between 1 and 2. The Sierpinski triangle contains three copies at $1/2$ scale, so has dimension

$$D = \frac{\log(3)}{\log(2)} = 1.5850$$

to four decimal places. Extending beyond two dimensions, the Sierpinski sponge has 20 self-similar copies at $1/3$ scale, giving dimension of

$$D = \frac{\log(20)}{\log(3)} = 2.7268$$

to four decimal places. This method works very well for objects which contain exact subcopies of themselves; but not very many objects are so well behaved, even if they have exact integer dimensions. Hausdorff's definition overcomes this problem. He uses the concept of a 'neighbourhood', which for our purposes can be considered as a small region of regular shape centred on a particular point. In normal one-dimensional space a neighbourhood is a short line segment, in two dimensions it is a small circle, in three a small sphere, each having the reference point at its centre. Imagine your object sitting in a particular space; for example, the Koch curve would sit within a two-dimensional space. Now cover the object with neighbourhoods of a particular size. Suppose it takes N_1 such neighbourhoods to cover it completely, that is, all points of the object lie within one of the N_1 neighbourhoods when they are properly arranged. Perform the same exercise again, this time with neighbourhoods scaled by a factor f. Suppose the number of neighbourhoods required this time is N_2. Clearly, the values of N_1 and N_2 depend in some way on the sizes of neighbourhoods and the way in which they fit in to the general shape and size of the object. There is no exact formula for the relationship between these values, but if the neighbourhoods are very small a limiting ratio for N_2 and N_1 depends on the factor f by which the neighbourhoods are scaled. Felix Hausdorff (1868–1942) defined the dimension of the object as D, where, in the limit for infinitesimally small neighbourhoods,

$$\frac{N_2}{N_1} = \left(\frac{1}{f}\right)^D$$

thus giving D as the limiting value of

$$D = \frac{\log(N_2/N_1)}{\log(1/f)}$$

This relates back directly to the discussion above for exactly self-replicating objects. Consider an object which has N copies of itself, each of scale factor f. If it takes N_1 neighbourhoods of a particular size to cover the object, then it also takes N_1 neighbourhoods at scale f to the original neighbourhoods to cover one copy of the object. Thus, the whole object can be covered by $N_2 = NN_1$ neighbourhoods which are f times the size of the original. Thus

$$D = \frac{\log(N_2/N_1)}{\log(1/f)} = \frac{\log(N)}{\log(1/f)}$$

as before.

To approximate the Hausdorff dimension or fractal dimension of a naturally occurring or synthetic fractal object, neighbourhoods can, in practice, be replaced by other small regular shapes such as squares and cubes. For example, to estimate the fractal dimension of a coastline trace a map of the coast, at two different scales, onto squared paper. For scale 1, count the number of squares which the coastline cuts through, N_1. For scale 2 let the number of squares occupied by the coastline be N_2. If scale 2 is f times larger than scale 1, then the size of the squares for map 2 has effectively been reduced by a factor f. Then the fractal dimension of the coastline can be estimated as

$$D = \frac{\log(N_2/N_1)}{\log(1/f)}$$

as above. If a number of different scales are used, provided the maps are of reasonable accuracy the values of D obtained for a set portion of fractal coastline are remarkably similar.

This has not been a rigorous mathematical development of the concept of fractal dimension; the discussion has been anecdotal and imprecise. However, it is hoped that the concept has been illustrated well enough to give the reader enough understanding of it to be able to follow the remaining papers.

Natural and Synthetic Fractals

Examples have been given in this paper of fractal objects generated by mathematicians and artists long before the word 'fractal' was coined. Fractal objects, however, predate all these man made constructs through their embodiments in natural history. The pattern of blood vessels in a human body is fractal, many naturally growing plants have always had fractal properties, coastlines and clouds can properly be considered as examples of fractal geometry [Voss88]. The fractal objects presented so far in this paper are deterministic. Although random selection has been used as a device in creating some examples, the fractals created do not vary; we have merely selected which points within the fractal sets to display in a random manner. The simulation of naturally occurring fractal objects involves building random variation into the fractal object itself. For example, in Pythagoras tree of Figure 19 it would have been relatively simple to incorporate a random perturbation of the angle of the triangle concerned at each repetition, thus creating an irregular ramifying fractal [Lauw87; Saup88; Saup90]. By such methods convincingly realistic objects can be created. The deceptively simple two-dimensional tree of Figure 22 (created by Aurelio Campa at Middlesex Polytechnic) can be synthesised into beautiful images such as Figure 23 and have been used to enhance otherwise cold architectural illustrations. A three-dimensional fractal tree (the work of Hugh Mallinder) is shown in Plate 1. Landscapes, mist and clouds have all been fractally generated by Semannia Luk-Cheung (Plates 2 and 3), with changes in fractal dimension enabling different forms of terrain to be modelled (see also [Voss88; McGu88]).

Conclusion

This has been a personally selective tour of some ideas in the theory of fractals. The development has not been rigorous—there are many good texts that fulfill the need for rigour (for example, Peitgen and Richter's book [Peit86] has several well documented mathematical sections, as well as many parts which are accessible to the layman). The purpose of the paper was to introduce a number of fractal objects to act as an introduction to the remaining sections of the book. Those who have not yet dabbled in the production of fractal objects should try

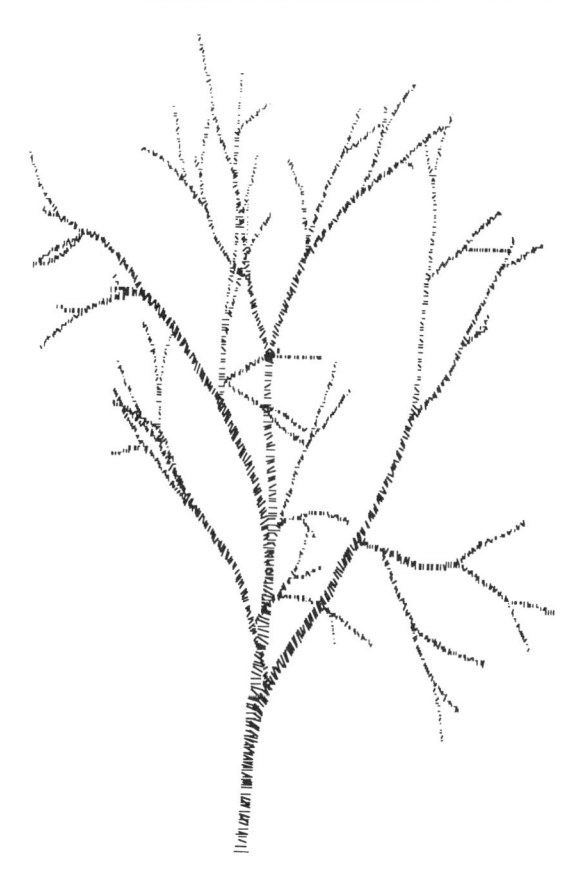

Fractals Before Mandelbrot—A Selective History 31

Figure 22. A two-dimensional fractal tree created by Aurelio Campa.

Figure 23. Aurelio Campa's misty forest.

their own variations on the methods outlined, to find out for themselves that fractals are fun, and fractals are fascinating.

Acknowledgements. The images accompanying this paper were created at Middlesex Polytechnic, using PICASO and PRISM software created by Professor John Vince, Paul Hughes and others at the institution. My thanks go to them, to the staff of the Middlesex Polytechnic Computer Centre and the Faculty of Engineering, Science and Mathematics for their support. I am also particularly indebted to Aurelio Campa, Hugh Mallinder and Semannia Luk-Cheung of the Faculty of Art and Design at Middlesex Polytechnic for permission to reproduce images created by them.

REFERENCES

[Abbo87]
Abbott, E.A., *Flatland, a Romance of Many Dimensions by A.Square*, Harmondsworth, UK: Penguin, 1987.

[Barn88a]
Barnsley, M.F., Fractal Modelling of Real World Images, in *The Science of Fractal Images*, Peitgen, H.-O., and Saupe, D., Eds., New York: Springer-Verlag, 1988.

[Barn88b]
Barnsley M.F., *Fractals Everywhere*, San Diego: Academic Press, 1988.

[Boye68]
Boyer, C.B., *A History of Mathematics*, New York: John Wiley and Sons, 1968.

[Cour69]
Courant, R., and Robbins, H., *What is Mathematics?* (4th ed.), Oxford, U.K.: Oxford Univ. Press, 1969.

[Cund61]
Cundy, H.M., and Rollett, A.P., *Mathematical Models* (2nd ed.), Oxford, UK: Oxford Univ. Press, 1961.

[Dixo87]
Dixon, R., *Mathographics*, Oxford, UK: Basil Blackwell, 1987.

[Doua86]
Douady, A., Julia Sets and the Mandelbrot Set, in *The Beauty of Fractals*, Peitgen, H.-O., and Richter P.H., Eds., Berlin: Springer-Verlag, 1986.

[Glei88]
Gleick, J., *Chaos*, Harmondsworth, UK: Penguin, 1988.

[Hofs80]
Hofstadter, D.R., *Gödel, Escher, Bach: An Eternal Golden Braid*, Harmondsworth, UK: Penguin, 1980.

[Kaan91]
Kaandorp, J.A., Modelling Growth Forms of Sponges with Fractal Techniques, in *this volume*, pp. 71—88, 1991.

[Lauw87]
Lauwerier, H.A., and Kaandorp, J.A., Tutorial: Fractals (Mathematics, Programming and Applications), Nyon, Switzerland: Eurographics, Amsterdam, 1987.

[Mand86]
Mandelbrot, B.B., Fractals and the Rebirth of Iteration Theory, in *The Beauty of Fractals*, Peitgen, H.-O., and Richter P.H., Eds., Berlin: Springer-Verlag, 1986.

[Mand88]
Mandelbrot, B., People and Events Behind the 'Science of Fractal Images', in *The Science of Fractal Images*, Peitgen, H.-O., and Saupe, D., Eds., New York: Springer-Verlag, 1988.

[McGu88]
McGuire, M., An Eye for Fractals, in *The Science of Fractal Images*, Peitgen, H.-O., and Saupe, D., Eds., New York: Springer-Verlag, 1988.

[Peit86]
Peitgen, H.-O., and Richter, P.H., Eds., *The Beauty of Fractals*, Berlin: Springer-Verlag, 1986.

[Pick86]
Pickover, C.A., Biomorphs: Computer displays of biological forms generated from mathematical feedback loops, *Comput. Graph. Forum*, Vol. 5, pp. 313–316, 1986.

[Pieg87]
Piegl, L., and Tiller, W., Curve and surface constructions using rational B-Splines, *Comput. Aid. Des.*, Vol. 19, pp. 485–498, 1987.

[Popp76]
Popper, K.R., *Unended Quest: An Intellectual Autobiography*, London: Fontana, 1976.

[Reev91]
Reeve, D.E., Mandelbrot, Julia Sets and Nonlinear Mappings, in *this volume*, pp. 35—42, 1991.

[Saup88]
Saupe, D., A Unified Approach to Fractal Curves and Plants, in *The Science of Fractal Images*, Peitgen, H.-O., and Saupe, D., Eds., New York: Springer-Verlag, 1988.

[Saup91]
Saupe, D., Random Fractals in Image Synthesis, in *this volume*, pp. 89—118, 1991.

[Voss88]
Voss, R.F., Fractals in Nature: From Characterisation to Simulation, in *The Science of Fractal Images*, Peitgen, H.-O., and Saupe, D., Eds., New York: Springer-Verlag, 1988.

Mandelbrot, Julia Sets and Nonlinear Mappings

Dominic E. Reeve

Abstract

Methods of computing Mandelbrot and Julia sets for a variety of nonlinear mappings are described. The original Mandelbrot set is constructed using a quadratic mapping. In this paper this is used as the first step in a numerical investigation of the properties of the Mandelbrot sets and the corresponding Julia sets for higher order mappings. The numerical results suggest several interesting relationships between the order of the mapping chosen and the rotational symmetries of the associated Mandelbrot and Julia sets.

Introduction

Mathematicians have been aware of pathological curves, or fractals, for many years. However, the publication of the first generally available book on the subject (by Mandelbrot [Mand77], who coined the term 'fractal') has provoked widespread interest in fractals. Among the first fractals to be drawn were those which had a regular geometric structure. Mandelbrot introduced the idea of random fractals in order to describe structures that occur in nature more realistically. An article by Batty [Batt85] gives several interesting examples of fractals in nature.

This paper returns to the original 'Mandelbrot Set', which is based on the repeated application of a quadratic mapping in the complex plane. Thorough details of this set and the corresponding Julia sets are given in Peitgen and Richter [Peit86].

In fact there are many Mandelbrot sets, any nonlinear mapping giving rise to a set with different features. This point is pursued further below.

Calculation of Mandelbrot Sets

The original Mandelbrot set, denoted by $M_2(z_0)$, is generated by a repeated mapping of the form $z \to z^2 + c$, where z and c are complex numbers. z_0 is the initial value of z and is usually taken to be $(0, 0)$.

For a chosen value $c = (p, q)$ the above mapping is applied repeatedly until one of two criteria is satisfied. These criteria are:

If after the maximum allowable number of iterations the magnitude of the iterate is less than a chosen 'critical magnitude', the point (p, q) is assigned the 'background' colour.

If the magnitude of the iterate exceeds the 'critical magnitude', the point (p, q) is assigned a colour according to the number of iterations that have already been executed.

The Mandelbrot set is constructed by performing this iterative procedure for a series of values of p and q. It may be interpreted as a map (in pq space) of the rate at which a given starting point, z_0, moves towards infinity under the mapping $z \to z^2 + c$.

In practice, one takes a finite set of values for $c = p + iq$ with $p \in [p_{min}, p_{max}]$ and $q \in [q_{min}, q_{max}]$. Having selected the domain of c, a method of discretisation must be chosen. This is done in the following way. The figure for $M_2(z_0)$ will be generated from the discrete values of c given by $c_{mn} = p_m + iq_n$, where $p_m = p_{min} + (m-1)\Delta p$, for $m = 1, 2 \ldots M$, and $q_n = q_{min} + (n-1)\Delta q$ for $n = 1, 2 \ldots N$, where M and N are the number of grid points in the chosen intervals for p and q, respectively. The increments are defined by $\Delta p = (p_{max} - p_{min})/(M - 1)$ and $\Delta q = (q_{max} - q_{min})/(N - 1)$.

The crucial part of the algorithm is the choice of p_{min}, p_{max}, q_{min}, q_{max}, the 'critical magnitude', and the maximum number of iterations that should be allowed. Values of these parameters which produce good figures have been given by Peitgen and Richter [Peit86]. A study of the sensitivity of the computed Mandelbrot set to changes in the critical magnitude and the maximum number of iterations (MAXIT) has been reported by Reeve [Reev89]. For a fixed mapping, different choices of z_0 will give rise to distorted versions of the 'canonical' set obtained for $z_0 = (0,0)$ [Dewd87]. For the Mandelbrot set M_2 a good set of parameter values is: $p_{min} = -2.25$, $p_{max} = 0.75$, $q_{min} = -1.5$, $q_{max} = 1.5$, MAXIT $= 200$ and critical magnitude $= 10$. In the computational algorithm the square of the critical magnitude is used in order to save repeated use of the SQRT function.

In the following black and white figures both M and N are set to 512. The computational time required is typically three quarters of an hour on a SUN 3/50 workstation.

Figure 1 shows the canonical Mandelbrot set $M_2(0,0)$ as calculated using the parameter settings given above. Figure 2 shows the Mandelbrot set calculated with the maximum number of iterations reduced to 10. The basic outline of the set remains, but all the interesting detail near the boundary of the set has been lost. In fact, as a general rule, the closer a point is to the Mandelbrot set the more iterations are required to determine its fate. This is the price to be paid for obtaining high resolution pictures of the set. As suggested by Peitgen and Richter [Peit86] one can choose the ranges of p and q to 'zoom in' on a particular portion of the Mandelbrot set. If the maximum number of iterations is not large

Figure 1. Mandelbrot set M_2 with MAXIT = 200, critical magnitude = 10. p runs from -2.25 to 0.75, and q runs from -1.5 to 1.5.

Figure 2. As Figure 1 but with a value of MAXIT = 10.

enough to allow the differentiation of the fate of z_0 for each value of c, one is likely to obtain a disappointingly bland figure. When computing such 'close up' figures one must expect to pay a price in substantially increased computational time due to raising the maximum number of iterations.

Calculation of Attractor Basins and Julia Sets

The Mandelbrot set is a map of the rate at which the point z_0 moves towards infinity under repeated application of a specified mapping. A closely linked idea is that of basins of attraction and Julia sets. These are calculated in a manner similar to the Mandelbrot set. In this case the value of c is fixed, and a map of the rate at which points in the complex plane move to infinity under the chosen mapping is plotted. As before, the same criteria are used to judge whether a point is in the attractor basin. That is, a domain of interest is defined by $x_k = x_{\min} + (k-1)\Delta x$, $y_l = y_{\min} + (l-1)\Delta y$, $k = 1, 2 \ldots K$, $l = 1, 2 \ldots L$ with $\Delta x = (x_{\max} - x_{\min})/(K-1)$ and $\Delta y = (y_{\min} - y_{\max})/(L-1)$. Then, each point $z_{kl} = (x_k, y_l)$ undergoes repeated mapping by $z_{kl} \to z_{kl}^2 + c$, until either the maximum number of iterations have been performed, in which case z_{kl} is taken to belong to the attractor basin whose boundary is the Julia set, or the iterate exceeds the maximum magnitude, in which case z_{kl} is coloured according to the number of iterations performed. In practice, z_{kl} will correspond to a vertex of a grid rectangle, and the whole rectangle will be coloured according to the fate of this vertex. Figure 3 shows the Julia set, denoted by $J_2(c)$, and attractor basin for the mapping $z \to z^2 + (0.5, 0.5)$, with $x, y \in [-1.5, 1.5]$. Peitgen and Richter [Peit86] discuss the connection between the Mandelbrot set and the Julia set for the mapping $z \to z^2 + c$. In essence, if c is outside M_2 then the corresponding

38   Dominic E. Reeve

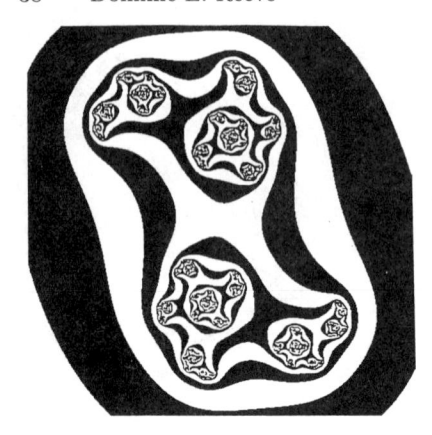

Figure 3. The attractor basin and Julia set, $J_2(0.5, 0.5)$. x and y run from -1.5 to 1.5.

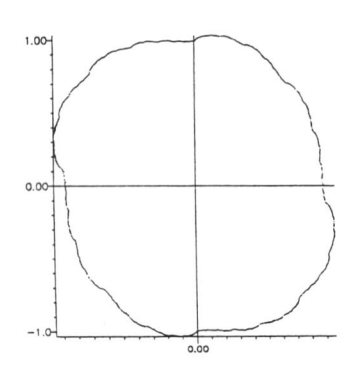

Figure 4. The outline of the Julia set $J_2(0.1, 0.1)$ using the inverse method. Both axes are ticked at intervals of 0.1.

Julia set will not be connected. Conversely, if c is within M_2, then the Julia set will be connected.

A useful, alternative method of obtaining the Julia set $J_2(c)$ has been given by Sorensen [Sore84] and involves repeated inversion of the mapping $z \to z^2 + c$. It provides a quick way of finding the appropriate bounds on x and y before one computes the attractor basin for a given value of c with the first method. Figures 4 to 8 illustrate the power of the method and the dependence of the Julia set on the value of c. The values of c are $(0.1, 0.1)$, $(0.3, 0.3)$, $(0.375, 0.375)$, $(0.4, 0.4)$ and $(0.5, 0.5)$, respectively, and all lie on the line $x = y$.

Referring back to Figure 1, for values of c near the origin the Julia set resembles a slightly deformed circle. As c moves away from the origin along the line xy the Julia set becomes more contorted, until it eventually breaks up into 'fractal dust'. In Figures 6, 7 and 8 the Julia set is no longer connected, as c does not

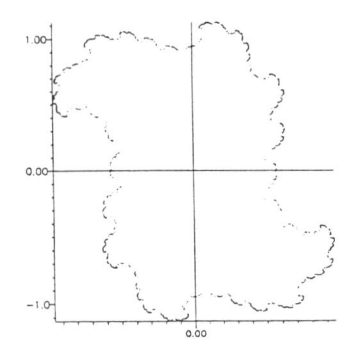

Figure 5. As Figure 4 but for $c = (0.3, 0.3)$.

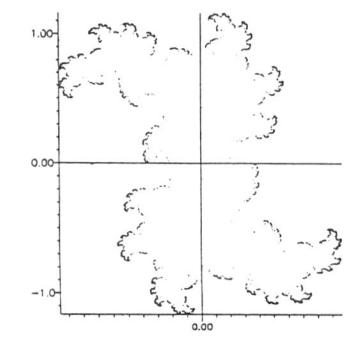

Figure 6. As Figure 4 but for $c = (0.375, 0.375)$.

Mandelbrot, Julia Sets and Nonlinear Mappings 39

Figure 7. As Figure 4 but for
$c = (0.4, 0.4)$.

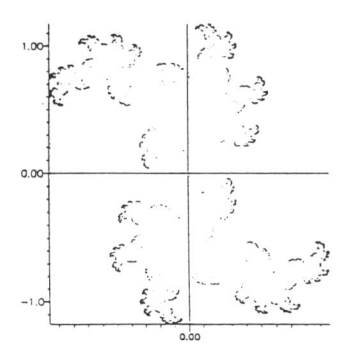

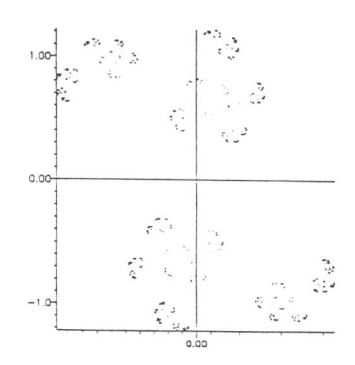

Figure 8. As Figure 4 but for
$c = (0.5, 0.5)$.

lie within M_2 in these cases. Axes have been drawn on Figures 4 to 8 for ease of reference, with tick marks at intervals of 0.1 on each axis.

More Mandelbrot and Julia Sets

The algorithms described above can easily be adapted to calculate the Mandelbrot and Julia sets (denoted by M_n and J_n, respectively) for the general mapping $z \to z^n + c$. For the present n is restricted to 3, 4 and 5, with $z_0 = (0,0)$.

Figures 9, 10 and 11 show the canonical Mandelbrot sets M_3, M_4 and M_5. All the parameter settings were the same as for the calculation of M_2 with the exception of the ranges of p and q, which were changed to $p, q \in [-1.5, 1.5]$. Together with Figure 1, these figures provide an intriguing sequence. M_2 has one axis of symmetry (about the line $p = 0$) and one main lobe off its main body. M_3 has two axes of symmetry (about the lines $p = 0$ and $q = 0$) and two main

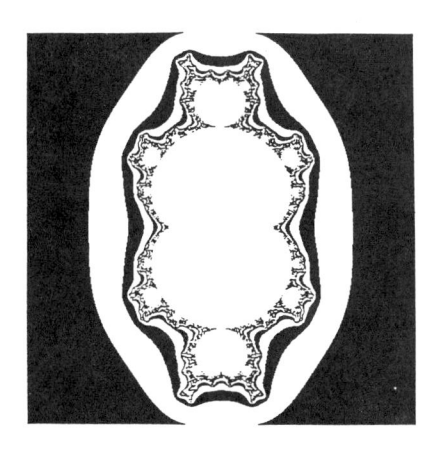

Figure 9. The Mandelbrot set M_3. p and q both run from -1.5 to 1.5.

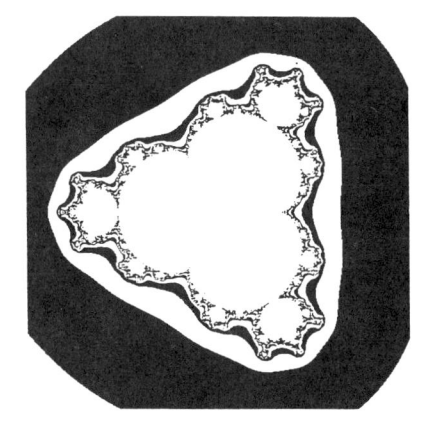

Figure 10. The Mandelbrot set M_4. p and q both run from -1.5 to 1.5.

Figure 11. The Mandelbrot set M_5. p and q both run from -1.5 to 1.5.

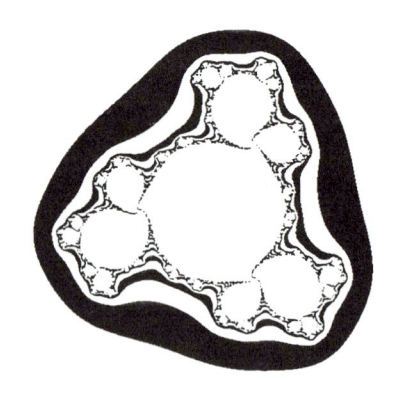

Figure 12. The attractor basin and Julia set $J_3(0.5, 0.5)$. x and y both run from -1.5 to 1.5.

lobes off its main body. M_4 has three axes of symmetry (about the lines $q = 0$, $q = \pm p\sqrt{3}$), and M_5 has four axes of symmetry (about the lines $p = 0$, $q = 0$, $p = q$ and $p = -q$). It should be emphasised that these 'symmetries' have been deduced on the basis of the figures rather than any analytical results. However, the pictorial evidence is very persuasive, especially in light of the figures for the corresponding Julia sets.

Figures 12, 13 and 14 show Julia sets $J_3(0.5, 0.5)$, $J_4(0.5, 0.5)$ and $J_5(0.5, 0.5)$. Together with Figure 3, these again provide a very interesting sequence. J_2 has rotational symmetry (about the origin) of order 2, while J_3, J_4 and J_5 have rotational symmetry of order 3, 4 and 5, respectively.

The rules governing the relationship between M_2 and J_2 discussed in Peitgen and Richter [Peit86] would seem to hold for the sets for higher order mappings. That is, for M_4 and M_5 the point $(0.5, 0.5)$ lies well within the main body of the Mandelbrot set, and the corresponding Julia sets are connected and have a generally smooth appearance. In total contrast, $J_3(0.5, 0.5)$ appears highly contorted though still connected. The reason for this dramatic change is that $(0.5, 0.5)$ does not lie in the main body of M_3 but lies in a secondary bud, close to the join with the main body. Finally, in the case of $J_2(0.5, 0.5)$, $(0.5, 0.5)$ lies outside M_2 and so is not connected.

On the basis of the figures produced by numerical computation, one is encouraged to make the following general postulates. First, M_n (corresponding to the mapping $z \to z^n + c(z_0 = 0, 0)$) will have $n - 1$ axes of symmetry, for n integer and $n > 1$. Secondly, $J_n(c)$ will have rotational symmetry of order n, for integer and $n > 1$. No analytical evidence for such results is attempted here.

Finally, with the benefit of hindsight, Plate 4 shows a colour version of M_3 over the region $p \in [-0.75, 0.75]$ and $q \in [-1.5, 1.5]$. This was computed using only the region defined by $p \in [0, 0.75]$, $q \in [0, 1.5]$ discretised with 63×126 points. Although this is approximately twice the grid point spacing used in the

Figure 13. The attractor basin and Julia set $J_4(0.5, 0.5)$. x and y both run from -1.5 to 1.5.

Figure 14. The attractor basin and Julia set $J_5(0.5, 0.5)$. x and y both run from -1.5 to 1.5.

black and white plots, the calculations took only five minutes to perform on a Compaq 286.

Discussion

The idea of Mandelbrot sets has been extended to cover mappings of orders larger than 2. Computer generated plots have uncovered what appears to be an extremely interesting relationship between the order of the mapping employed and the symmetries of the Mandelbrot and Julia sets. As has been demonstrated, these fascinating and intricate figures can be computed without resorting to vast computer resources. The opportunity to explore the beauty of these sets is now available to anyone with access to a PC.

Acknowledgements. Figures 1 to 14 are reproduced with the kind permission of Marconi Underwater Systems Ltd.

REFERENCES

[Batt85]
   Batty, M., Fractals—geometry between dimensions, *New Scient.*, pp. 31–35, April 1985.

[Davi87]
   Davies, I.M., Space filling curves and fractals on micros, *IMA Bulletin*, Vol. 23, pp. 94–99, 1987.

[Dewd87]
   Dewdney, A., Beauty and profundity: the Mandelbrot set and a flock of its cousins called Julia, *Sci. Amer.*, Vol. 257, No. 5, pp. 140–145, 1987.

[Mand77]
Mandelbrot, B., *Fractals: Form, Chance and Dimension*, San Francisco: W.H. Freeman, 1977.

[Peit86]
Peitgen, H.-O., and Richter, P.H., Eds., *The Beauty of Fractals*, Berlin: Springer-Verlag, 1986.

[Reev89]
Reeve, D.E., Some relations of the Mandelbrot and Julia sets: a computational exploration, *IMA Bulletin*, Vol. 25, pp. 185–191, 1989.

[Sore84]
Sorensen, P., Fractals, *Byte*, pp. 157–172, Sept. 1984.

Cities as Fractals: Simulating Growth and Form

Michael Batty

Abstract

The morphology of cities bears an uncanny resemblance to those dendritic clusters of particles which have been recently simulated as fractal growth processes. This paper explores this analogy, first presenting both deterministic and stochastic models of fractal growth, and then suggesting how these models might form an appropriate baseline for models of urban growth. In particular, the diffusion limited aggregation (DLA) and dielectric breakdown models (DBM) are outlined, and comparisons are made between simulated clusters and the form of medium-sized towns. Using the DBM formulation, constraints and distortions on the simulated clusters are illustrated, thus enabling the model to simulate a continuum of cluster forms from the one- to the two-dimensional. This model is then applied to the town of Cardiff, UK, and a critical evaluation of its performance is used to identify future research.

Introduction

The best known physical example of a fractal is a coastline. Mandelbrot, who is accredited with the definition, rekindled interest in the conundrum [Mand67] as to the exact length of a coastline which was posed and illustrated by Richardson [Rich61] almost 30 years ago. Richardson demonstrated quite unequivocally that the length of a coastline depended upon the yardstick or scale with which its length was measured. He showed that as the scale became finer more and more detail was picked up by the measuring instrument, thus implying that there were no bounds upon its length. Mandelbrot [Mand67] took the analysis further in pointing out that coastlines were simply one from a much larger class of objects in which length in one dimension had no meaning. Lines which were 'more than straight' filled the available space to a degree which could be characterised by a fractional (hence fractal) dimension, greater than the one dimension of the line but less than the two dimensions of the plane.

This conundrum of length has in fact been known for a long time. In the 1960s, the Michigan Inter-University Community of Mathematical Geographers was centrally involved in measuring the properties of such geographic space. In a seminal paper Nysteun [Nyst66] not only identified the problem and suggested the definition of length contingent upon the scale used, but he also pointed to the work of Steinhaus [Stei54, Stei60] and to the Polish geographer Perkal [Perk58a, Perk58b], who in turn noted that the Viennese geographer Albrecht Penck [Penc94] had been familiar with the problem in the late nineteenth century. There is some evidence too that Leonardo da Vinci had wrestled with it, and if Leonardo knew about it, so must have the Greeks.

What is fascinating about the problem is that it has never been restricted to physical systems. Nysteun [Nyst66] described boundary problems in the town of Ann Arbor which displayed the conundrum, Richardson [Rich61] used frontiers as examples, and Perkal [Perk58b] illustrated methods for resolving the question using the boundary of the city of Wroclaw as an example. Nor does Mandelbrot consider that fractals should be restricted to physical problems. He says, in discussing the amount of circuitry which can be packed onto a chip: 'This and a few other case studies help demonstrate that in the final analysis, fractal methods can serve to analyze any 'system', whether natural or artificial, that decomposes into parts in a self-similar fashion, and such that the properties of the parts are less important than the rules of the articulation' [Mand82, page 114].

The second strand of work which has long been of central concern to those dealing with the spatial structure of cities involves the idea of self-similarity. Coastlines and man-made boundaries, of course, show self-similarity; but the distribution of cities and their arrangement as central places, in terms of both size and spacing, illustrate distinct hierarchical ordering. Christaller's central place theory [Chri66], Zipf's rank size rule for cities [Zipf49], as well as the many analogies between city and physical systems promoted by Stewart and Warntz [Stew58] amongst others as 'social physics', have dominated the development of urban theory and regional science since the 1950s. Although this concern has not emphasised the morphology of city systems *per se*, there has been a stream of *ad hoc* research concerned with spatial form which is consistent with social physics on the one hand and the economic operation of cities on the other.

This concern for form is best seen in the many analogies in geography between physical and human systems—rivers and central places, for example [Wold67], as well as in various attempts to research the growth of cities using allometry [Dutt73, Wold73]. Indeed, in the late 1960s there was a group of researchers calling themselves the 'Philomorphs', who met at Harvard and whose concern was primarily to develop morphological analogies between physical and human systems. This group, amongst whose members were Stephen Gould, Peter Stevens, Cyril Smith and Michael Woldenberg, were centrally concerned with the study of form. A clear expression of their interest is contained in Stevens' book *Patterns in Nature* [Stev74]. Much of this work clearly predated the development of fractals, although their objects of study were essentially fractals. It does demonstrate, however, the concern amongst researchers in the human and social as well as the physical sciences for systematic order in the morphology of artifacts.

Since the popularisation of fractals during the last decade, a number of the themes in urban geography and regional science noted above have been picked up and developed with renewed vigour. The use of computer graphics in simulating the irregular form of cities in terms of land use development [Batt86], work on map generalisation and cartography using fractal ideas (for example, see [Butt85, Mull86, Mull87]), research into how central places can be generated through their hierarchy using deterministic fractals [Arli85], the measurement of urban boundaries and edges [Batt88]—these are but a few of the applications of fractal geometry to city shape, order and form. A useful review is given by Goodchild and Mark [Good87] who contrast applications in both physical and human geography, and it is already clear that during the next ten years these various applications will converge and enrich the traditional systematic study of cities.

Developments of fractal geometry in this domain are not simply the indulgence of pure research. There is, in the everyday world of the market place and of government, a strong interest in the development of better methods for simulating, predicting and measuring the properties of urban form. Classification of cities according to their various shapes has always been an important influence on the way cities are planned and conserved. Cities are frequently conceived in terms of simple Euclidean geometry, as seen for example in the linear and concentric forms of the ideal city proposed by Le Corbusier and other such eminent architectural practitioners. The planning system as it has grown up in the West is based upon guiding and manipulating physical change in terms of spatial, hence geometric policy instruments ranging from new towns to the containment of urban growth. Plans for the future are often based upon ways of concentrating growth or spreading it out into the suburbs, and it is thus clear that the study of urban form must be central to such concerns. In this milieu it is certain that fractal geometry will have something of importance to contribute, if only in sharpening new questions or in providing insights into longstanding ones.

There is another reason why fractal geometry must be of concern to urban researchers and city planners. An arsenal of sophisticated techniques for analysing and predicting urban structure has been assembled over the last decade, but their relation to urban form is at best tentative. Models of economic and demographic activity in cities are largely based on representing spatial form at a level of abstraction from which the city's geometry cannot be easily construed. Models are built on data at the level of the census tract, while others either do not attempt to represent the spatial dimension whatsoever or define it away using simplistic and often inappropriate conceptions of Euclidean geometry. Thus the types of fractal geometry we will suggest in this paper lead to models of urban structure which are in no way alternatives to contemporary practice, but which are in every sense complementary to existing approaches.

What we will do here is outline how fractal processes can generate highly ordered clusters of particles in two-dimensional space, and we suggest that the methods and models of these processes recently developed for physical systems might constitute useful analogies for city growth. The processes in question generate clusters which are far-from-equilibrium, in that their growth processes are irreversible. The sorts of clusters produced are tree-like structures—dendrites— which have evident self-similarity in their branching, and which apply to a range of physical systems all driven in some way by diffusion of particles from some source. For example, physical growth such as electro-deposition, viscous fingering, crystallisation, dielectric breakdown and various forms of percolation have all been recently simulated using fractal growth processes. A recent review is provided by Vicsek [Vics89], but there has been a veritable explosion of applications since the early 1980s, when the original suggestion by Witten and Sander [Witt81] of a simple model of a diffusing particle whose behaviour could be simulated as a random walk on a two-dimensional lattice was made. This model is based on the aggregation of particles, one at a time, whose diffusion is limited or constrained by a fixed field of influence around the growing cluster, and by the fact that once a particle reaches the cluster it sticks permanently. This model, called, appropriately enough, 'Diffusion Limited Aggregation' or the DLA model, has been widely researched; a recent review of the current research questions under scrutiny is given by Stanley [Stan87a], while a popular account is provided by Sander [Sand87].

In the rest of this paper we explore how the DLA model might be adapted to the simulation of city growth and form. In the next section we examine the typical form of some large cities and present a deterministic model of growth which generates highly regular but sparse fractal structures which have physical properties with some similarity to the structure of cities. In short, we suggest an idealised generator of form and show how its fractal dimension might be measured, with the implication that the same sort of process might be used to simulate the form of cities. We then introduce two models of fractal growth which incorporate a degree of randomness in their growth processes—the DLA model, and the Dielectric Breakdown Model or DBM, both of which are part of a class of processes which are based on 'Gradient Governed Growth' (or GGG). We then show how their fractal dimension and other properties can be measured and estimated, and we outline how their growth processes can be distorted to produce a continuum of geometrical forms over a range of fractal dimension. In applying these ideas to real cities, it is necessary to deal with irregular and distorted space fields in which growth takes place. Thus, by demonstrating this theoretically we are then in a position to apply such ideas to real cities. We have chosen the example of the medium-sized town of Cardiff (population approximately 302,000 in 1981) as the testbed for our simulations using the DLA-DBM model, and we tentatively suggest how such simulations might be developed. Our conclusion emphasizes that although our analogies may be somewhat speculative, this approach appears to have considerable promise in providing a shell within which more complex simulations of other related urban phenomena might be embedded.

Urban Growth and Form

Scale, Size and Morphology

The form of cities can be visualized in very different ways at many levels of abstraction. It is therefore necessary to be completely clear as to the type of form in question. In Figure 1 we have illustrated the urban development and the urban hinterlands or fields of three large cities—London, Paris and Tokyo (in Figures 1a, 1b and 1c, respectively)— all at the same physical scale. These maps are adapted from those published by Doxiadis [Doxi68] and, although somewhat dated in relating to development in the early 1960s, the commonality of physical scale enables us to make immediate comparisons. In Figure 1d we show a

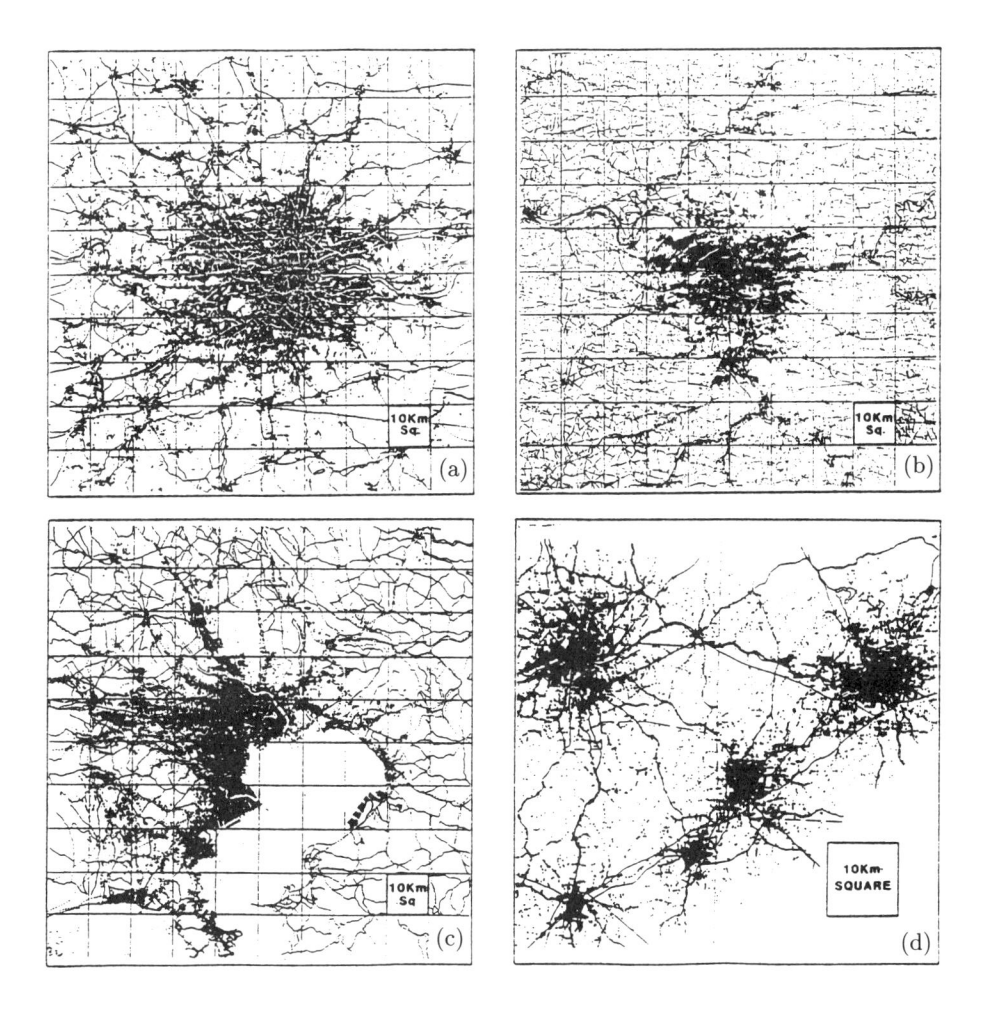

Figure 1. The morphology of large cities.

constellation of cities at a somewhat larger scale in the region of Greensboro, North Carolina, taken from Chapin and Weiss [Chap62]. This shows how cities grow into each other, as well as the fact that cities of different sizes at different scales have clear self-similarity of form when all four examples in Figures 1a to 1d are compared.

In examining the forms in Figure 1, it is clearly possible to detect tentacles of development emanating from the central cores of the cities in question. These tentacles have a dendritic form in that they follow the main lines of transport or other forms of communication from the core, or Central Business District (CBD), to the suburbs. It is through the provision of transport infrastructure radiating out from the core that the city's suburbs are opened up, and most Western cities which rely heavily on the automobile for personal transport are characterised by such strip or ribbon development. In one sense it is the transportation system which has the fractal structure of dendrites on which the rest of the city's development is hung. In the examples in Figure 1 there is a degree of self-similarity, in that the dendritic skeleton appears at different scales, as can be seen by comparing Figures 1a, 1b, and 1c with Figure 1d. However, what cannot be seen from these illustrations is the self-similarity of the way in which the city itself, its districts and its neighbourhoods are configured with the same forms of commercial and transportation structure.

Self-similarity of form and function usually exists over a restricted number of scales. This is very clear in terms of urban form. At physical scales finer than the neighbourhood, the geometry of form becomes Euclidean in that streets and dwellings and other physical structures are usually laid out following simple straight line geometry. For example, Hausmann's Paris of the mid-nineteenth century, the British New Towns, Roman military towns or castra—the examples are many throughout history. At the local scale all these are outside the scale range where fractal geometry might apply, although even here there may be deterministic fractal generators which are able to produce highly ordered layouts. Nevertheless, for our purposes the range over which we might expect the geometry of settlements to be irregular and self-similar probably begins at the district level and continues up to the regional, possibly the national and even the continental scale with respect to the type of form being measured and modelled.

We also need to be clear about the way city growth is handled in contemporary urban theory. It is usually assumed that cities can be examined with respect to forces which both 'pull' activity into them from their external environment, and forces which 'push' activity out from the core to the periphery. Moreover, both these centralising and decentralising forces are likely to be present to various degrees in any city . There are models which postulate city growth as some type of balance between the forces which attract migrants from the countryside or other cities and those which force newcomers or newly evolving households from the centre of the city to the edge. Moreover, there is the simple but relevant model which suggests that new growth must always occur at the edge of a city, because physical structures last much longer than the time it takes for migrants to arrive at the city and seek permanent accommodation. Thus the city mainly grows on its periphery; in cities where there is very rapid growth, peripheral growth accounts for almost all the new development.

Thus, any city at a fixed point in time represents a balance between the demand for space for growing activities which can most easily be accommodated on the edge or at a distance from the city, and the need to be near the core so that economies of scale are maximised. The cost of land and structures is highest at the core, where there is fierce competition for space; but the cost of transport to the periphery also acts as a deterrent to growth on the boundary. This is the sort of balance which the economic theory of cities attempts to simulate in predicting the size, type and location of economic activities in the city. The balance, then, is one between accessibility and economy which is maximised at the centre of the city, and the need for space and low costs of land which can best be met at distances further and further from the city. It is this balance that is also modelled by the fractal growth processes which we outline in the next main section.

What we require then is a model in which the potential for location with respect to space is at a maximum at further and further distances from the city, due to the fact that the cost of space is highest in the core and the availability of space is greatest at a large distance from the core. The precise balance depends upon many contextual factors relating to transport technology, the system of government practised, the economic wealth of the city and its nation state, and suchlike factors which strike the balance between push and pull effects. Before we look at these processes, however, we need to examine the way in which ordered physical forms within a given space might evolve in a theoretical manner, so that we can develop our methods with clarity before we apply them to more complex situations.

IDEALISED GENERATORS OF URBAN GROWTH

It is quite clear from Figure 1 that cities do not grow in dense compact clusters; this immediately suggests that their fractal dimension lies between 1 and 2. In short, the dendrites or fingers of growth which extend out into the countryside or urban field screen the space between them from receiving further growth, at least from the large amounts of growth which the tips of the fingers themselves attract. In this context, we need to be clear about what constitutes development; this includes all built structures which at any particular instant in time require complex demolition or institutional change for the space they occupy to be released for new growth. Thus we envisage urban development to be irregular, in that the urban fabric is likely to be peppered with holes or sites which cannot be and have not been developed so far. The model we now propose meets these requirements.

Following Vicsek [Vics89], we first examine the case of a static structure which has already been grown. We examine its self-similarity and compute its fractal dimension before we look at its complement, the growing fractal. In Figure 2a we show such a structure across four levels of finer and finer detail or scale. We can

50   Michael Batty

imagine that the instrument we have at our disposal for detecting the structure can be magnified to these finer scales and is thus able to see more detail. In Figure 2a, as we go down scale we see that the pattern at each scale repeats itself and is clearly self-similar. Moreover, although it fills space it does not fill the whole of the space available. The pattern is based on a regular generator in which the units of space at one level k contain five units of development at the next scale down, $k+1$. If we assume that the structure has a one-dimensional length scale L which is the side of the square at scale $k = 0$, then the unit of space at scale k, ξ_k is given as

$$\xi_k = \frac{L}{n_k} \qquad (1)$$

where n_k is the number of equal unit lengths into which L is divided. Clearly from Figure 2a, where $k = 0$, $n_k = 1$; where $k = 1$, $n_k = 3$; where $k = 2$, $n_k = 9$ and so on. In short, $n_k = 3^k$, and this represents the regular pattern of subdivision. Therefore, at each level k the total number of units into which the area L^2 is divided is n_k^2. In fact, the number of squares which are occupied or developed is given as N_k, which is less than n_k^2. Thus the number of developed units can be written as

$$N_k = n_k^D \qquad (2)$$

where D is the fractal dimension and $1 < D < 2$. This relation is usually written explicitly in terms of the scale ξ_k as

$$N_k = \left(\frac{L}{\xi_k}\right)^D = G\,\xi_k^{-D} \qquad (3)$$

where the constant of proportionality $G = L^D$. If the fractal is deterministic

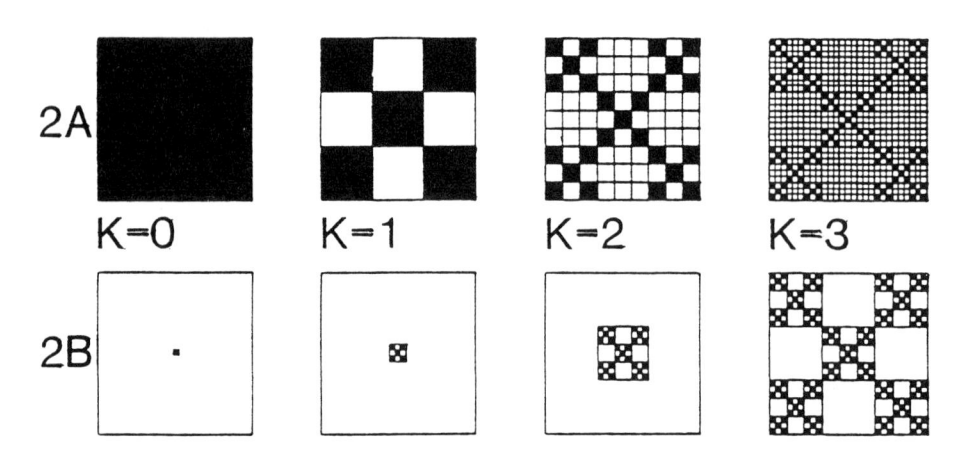

Figure 2. Scale and aggregation in static and growing fractals.

hence regular at each scale, as this object clearly is over the range $k = 0, 1, \cdots, K$, then from Eqs. (2) and (3) the fractal dimension can be written directly as

$$D = -\lim_{k\to\infty} \frac{\log N_k - \log G}{\log \xi_k} = -\lim_{k\to\infty} \frac{\log N_k}{\log \xi_k - \log L} \qquad (4)$$

This reduces to

$$D = -\lim_{k\to\infty} \frac{\log N_k}{\log \xi_k} \qquad (5)$$

if L, which is arbitrary, is set equal to 1, which is the usual convention.

For the regular fractal in Figure 2a, as $N_k = 5^k$ and $\xi_k = 3^{-k}$, substitution into Eq. (5) gives $D = \log(5)/\log(3) \approx 1.465$, which is the fractal dimension, a measure of the extent to which the fractal in Figure 2a fills its available space. What this effectively means is that although the number of units comprising the square object increases as the square of the scale, the total number of units occupied or 'developed' increases at less than this, as the fractal dimension D, which is less than the square but greater than the unit power. This implies that the density of occupation in fact decreases as the overall size of the fractal increases, a point we will take up a little later.

The method just outlined would be quite acceptable for determining the dimension of existing cities such as those shown in Figure 1. But it would have to be adapted to incorporate some statistical estimation of D across a range of scales, due to the fact that the self-similarity in such fractals would vary randomly within some range. But to examine the dimension of a growing object, the logic already developed can be extended in an alternate although complementary way. In Figure 2b we show the development of a fractal object across four scales, from a unit of development when $k = 0$ to a structure some 27 times as large when $k = 3$. In Figure 2b we grow the fractal from its basic seed, which now has a linear dimension ξ, to a cluster of seeds which spans the whole of the fixed space with linear dimension L, within which growth takes place. This is an idealised model of how a city might grow. The basic seed where $k = 0$ is the central place around which the city begins to grow. When $k = 1$, the seed spawns four outriders—say neighbourhoods—which constitute the basic unit of development. At $k = 2$ these modules are completely replicated around the basic module of $k = 1$, while when $k = 3$ the whole constellation of modules at $k = 2$ is repeated around the structure so far. Clearly the structure can continue to grow, but we have terminated it at $k = 3$ so that it becomes identical to the static structure in Figure 2a at this stage of its growth. It is important to note that k does not represent a real time scale but only a spatial one, although in some contexts it may be appropriate to think of k as embodying both space and time.

Formally we can define the number of units comprising the linear dimension of the cluster as n_k and L_k as the linear scale of the cluster grown up to k. Then

$$n_k = \frac{L_k}{\xi} \qquad (6)$$

or
$$L_k = n_k \xi \qquad (7)$$

Now, from Figure 2b it is clear that the total number of units N_k comprising the structure increases at less than the square of its dimension L_k but greater than unity. Assuming the growth power is D, then N_k is given as

$$N_k = L_k^D = (n_k \xi)^D \qquad (8)$$

As we have assumed that ξ is constant and equal to 1, Eq. (8) is then identical to Eq. (2). A simple manipulation of Eq. (8) yields Eq. (3), and it is thus clear that the fractal dimension D is the same as that defined in Eqs. (4) and (5). Figure 2 confirms this geometrically but with the assumption that the physical scale of N_k is different, according to whether the analysis is for a static or growing fractal.

Finally, we look at the density of occupation of the growing fractal, although the analysis is of course identical for the static case. We first assume that n_k is some measure of radial distance within the cluster defined as R_k, and that the area of the space within which the cluster is growing at any stage k is proportional to $R_k^2 = n_k^2$. The density of occupation defined as ρ_k is the ratio of N_k to R_k^2 and, assuming that $\xi = 1$ as above, the density is

$$\rho_k = \frac{N_k}{R_k^2} = R_k^{D-2} \qquad (9)$$

Equation (9) implies that the density decreases as the cluster grows larger. If R_k is a measure of radial distance, then the density around the seed site falls off with increasing distance away. This is the sort of physics one might expect in a growing city, and we see that Eq. (9) can easily be generalised to deal with a spatial system showing a growth pattern similar to that in Figure 2b but with the precise development configured randomly.

There is one last point to make which is clearly shown by the deterministic model described here. The growth process is irreversible, in that once a unit of development has been located there is no process whereby this can change. In the cities of the West, new growth can account for as little as 20 percent of the change per annum in locational patterns, although if only new physical development is considered then this proportion can be as large as 90 percent. Nevertheless, our model is extremely simplistic with respect to this assumption. Although at the level of detail shown in the urban patterns in Figure 1 the major changes in these cities over time would appear as new growth, more detailed changes within the fabric of the city generated through its social and economic processes cannot be handled by this type of theory. Extending the model to incorporate reversibility would not be difficult, and this remains a possibility for future research. But at the present, more insights into urban processes are likely to emerge if the model is kept in its most simple form.

Fractal Growth Processes

DIFFUSION-LIMITED AGGREGATION: THE DLA MODEL

A decade or so ago, Witten and Sander [Witt81, Witt83] suggested an elementary stochastic model which reproduced exceedingly rich, complex but nevertheless ordered, dendritic clusters of particles formed using simple principles of diffusion and aggregation. A typical system of interest is based on a square two-dimensional lattice, such as that implied in Figure 2 in which a seed is planted at the centre. At a large distance from this central seed a particle is launched and begins a random walk to adjacent lattice sites, one step at a time, entering the field of influence around the growing cluster. Once the field of influence (which may or may not be coincident with the launch site zone) is entered, the particle is either destroyed if it leaves the field or attached to the growing cluster when it reaches a lattice site at the edge of the cluster.

In this way the cluster grows, until a certain size is reached or a threshold is crossed. The resulting cluster is dendritic; it is obviously self-similar and thus is a fractal. At first glance the process of growth might seem to favour a compactly growing form, but what essentially happens is this: when the seed is first planted each of its adjacent sites has an equal probability of occupation, although only one of these is eventually chosen. Once this choice is made, the sites adjacent to the one chosen now have a slightly higher probability of being chosen. As the process continues, the cluster grows out in dendrites whose tips have a much stronger probability of being chosen than those sites which are located in the crevasses between. The tips of the cluster effectively screen the crevasses from further growth, and the cluster becomes increasingly tree-like as it continues to grow. The sort of structure which results is illustrated in Figure 3, where the process has been simulated on a 300 × 300 lattice, and where something over 4000 particles have been clustered.

The structure in Figure 3 has very obvious fractal properties which are similar to the deterministic structures in Figure 2. Firstly, there is self-similarity in

Figure 3. A simulated cluster using DLA or DBM.

the cluster's branches across several scales. Secondly, although the cumulative number of particles increases as the distance from the central seed increases, this does not increase as fast as the area contained in the field defined using this distance as a radius. In short, the density of the cluster falls as it gets bigger, as was also suggested in the case of the deterministic fractal of the previous section. Moreover, in analogy to Eq. (9), the number of particles at radial distance R from the seed site, given as $N(R)$, scales as

$$N(R) = GR^D \tag{10}$$

and the density scales as

$$\rho(R) = ZR^{D-2} \tag{11}$$

G and Z are constants of proportionality, and D, the fractal dimension, must be less than 2 but greater than 1. Witten and Sander's work [Witt81] has both stimulated and revealed an enormous amount of research into such processes. One of the most surprising findings is that the fractal dimension D appears to be approximately 1.71 ± 0.02, and this has been shown in countless cases although only for lattices up to about 1000×1000 in size. Recent work, in fact, has revealed that the dimension may still depend upon the size of the lattice and the number of particles grown in the cluster. Meakin [Meak86] has shown that as the lattice becomes larger a weak anisotropy begins to make itself felt due to the fact that the simulation is biased towards the occupation of sites at the four points of the compass on a square lattice. There is also some speculation that for mega-DLA simulations of upwards of one million particles, the dimension could well fall towards unity as the system increasingly forms a cross based on four all-dominant dendrites.

There has also been much work on extending these simulations to three and greater numbers of dimensions [Meak83a, Meak83b]. Some attempts at a field theory for such structures suggest that the dimension D might be related to the dimension of the lattice space d in a simple way. For example, Muthukumar [Muth83] suggests that the fractal dimension $D = (d^2 + 1)/(d + 1)$, which for a two-dimensional system gives a value of $D = 1.66 \cdots$ However, what is significant is not the precise value of this dimension but that the particles comprising the cluster and their density scale in a simple way with this dimension. Moreover, the dimension is not uniquely associated with any particular pattern. As Figure 2 so cogently illustrates, as long as the number of developed sites scales at less than the square of the units defining the linear dimension, the fractal dimension will be between 1 and 2.

In the case of Figure 2, the number of developed sites at any scale k is 5^k, which is more than the linear dimension 3^k and less than its square 3^{k^2}. This gives a fractal dimension of $D = \log(5)/\log(3)$, but this is not associated with any particular pattern, only one in which the scale grid increases in the order of 3 and the number of developed sites in the order of 5. These five units might be configured in several ways, and of course when random fractals are simulated as in Figure 3 an enormous number of possible configurations give the same fractal dimension. This has recently been widely recognised in the notion that different areas of the cluster scale differently, thus implying the existence of a very large number of fractal dimensions—multifractals, as they have been called. Recent research is reviewed by Stanley [Stan87a] and by Stanley and Meakin [Stan88]. Useful reviews which put the DLA model in context are presented in Feder [Fede88] and Vicsek [Vics89].

DIELECTRIC BREAKDOWN: THE DBM MODEL

Although it has been difficult to develop a consistent field theory for the growth of fractal clusters, the DLA model as originally formulated by Witten and Sander [Witt81] obviously incorporates diffusion. The growth problem can thus be treated as a problem in potential theory, in which the growth of the cluster seeks to maximise a potential but in an incremental, local not global, fashion, and subject of course to boundary constraints posed by the source of the diffusion and the growing cluster itself. In our analogy with the growth of cities, potential can be regarded as a measure of accessibility to available space. All other things being equal, as the distance from the edge of the cluster increases towards the edge of the field the potential based on the space available for development increases, until at a sufficiently large distance from the cluster it is assumed to be at a constant maximum. If the potential is defined as $\phi(x,y)$, where (x,y) are the coordinates of location, there are two boundary constraints of significance, namely $\phi(x,y) = 1$, which is an arbitrary constant at all locations beyond the edge of the field, and $\phi(x,y) = 0$, in and on the edge of the cluster. This latter constraint simply incorporates the notion that the process is irreversible.

Since the diffusion of particles in such a system is very slow, we can assume that the potential field is everywhere governed by Laplace's equation, i.e., that

$$\frac{\partial^2 \phi(x,y)}{\partial x^2} + \frac{\partial^2 \phi(x,y)}{\partial y^2} = 0 \qquad (12)$$

At any point in time the sites, which are adjacent to the cluster grown so far and are thus candidates for growth, have a probability of growth defined as $p(x,y)$ which is proportional to the gradient of the potential field, that is

$$p(x,y) \propto \frac{\partial \phi(x,y)}{\partial x} + \frac{\partial \phi(x,y)}{\partial y} \qquad (13)$$

As the potential on the boundary is zero, the gradient in Eq. (13) reduces to

$$p(x,y) \propto \phi(x,y) \qquad (14)$$

The growth of the cluster now depends upon the probability distribution associated with these boundary sites, for these sites are the only candidates for growth. Thus growth occurs by choosing the site with the greatest available space potential but making the choice randomly from the distribution given in

Eq. (14). In this way noise is introduced into the system, and the cluster develops its characteristic dendritic form. Although available space potential tends to be maximised, this is only in the immediate vicinity of the cluster, for the structure must remain connected—new sites must be physically attached—for the cluster to maintain the economies of scale associated with a city. In this way, a balance is struck between the desire to be as far away from the city as possible and the need to be within it.

This model has been used to simulate a variety of far-from-equilibrium real systems, in particular viscous fingering where a fluid of low viscosity permeates one with a higher viscosity (for example, water permeating oil), and dielectric breakdown, where a charge is released from some source and is attracted towards a more distant sink of high potential. The way the model is operated involves solving for the potential each time a particle is to be added to the growing cluster and then allocating the particle randomly to one of the perimeter cluster sites, the random allocation being based on the probability distribution $\{p(x,y)\}$. The way growth occurs would appear to be opposite to that of the DLA model, in that in DBM the growth develops from the centre out, whereas in DLA the growth is formed by the movement of particles from the periphery of the field towards the centre of the cluster. In these terms the DBM approach would appear to be more characteristic of city growth, although the process in both approaches is one of balancing push with pull. Niemeyer, Pietronero and Wiesmann [Niem84] provided the first and most complete model in this form for the dielectric breakdown problem, while Nittman, Daccord and Stanley [Nitt85] have developed several variants of the model for viscous fingering. The DBM approach can be considered as part of a more general approach to diffusion termed by Sherwood and Nittman [Sher86] as 'Gradient Governed Growth'; a formal review of these Laplacian fractals is provided by Stanley [Stan87b].

So far we have introduced three models of cluster formation, the deterministic fractal and the two analogous processes DLA and DBM, which both produce statistical or stochastic fractals. The tree-like forms produced by these processes have been widely acclaimed as being very close to the form of many physical systems, although it is clear in the study of city form that there are many types of cities which are both denser and sparser than the clusters produced by fractal growth. For example, cities in the western United States are much more spread out than those in the east and in Europe, where cities are older and have developed more compactly about their historical urban cores. However, it is possible to manipulate both the DLA and DBM approaches so that sparser and denser forms result, thus enabling different classes of form to be generated across the continuum from the line to the plane.

The dendritic structures produced by the DLA model essentially result from noise in the system due to the random walking of incoming particles. Such structures can be made more compact by reducing this noise or even sparser by increasing it. There are various approaches, all based on adding a particle to the cluster after it has reached the site in question a minimum number of times. For example, to generate a dense compact structure a particle only sticks when it has visited a peripheral site say 50 times. Thus, this enables the cluster to grow more compactly as the probabilities of occupation are effectively smoothed out. It is perhaps easier to see how this effect can be incorporated into the DBM model, where the noise in the system is caused by the random choice of a site according to the probability distribution of all boundary sites. The probabilities can be made larger even if the range of potential values $\{\phi(x,y)\}$ is reduced, and this can be done by scaling these to a power μ which is less than 1. If this parameter is greater than 1 then the range of probabilities will be increased, in that the more probable sites will dominate. The probabilities are now calculated from

$$p(x,y) \propto \phi(x,y)^{\mu} \qquad (15)$$

When $\mu = 1$ in Eq. (15), we retrieve the DBM model which generates the fractal case with $D \approx 1.71$, while as $\mu \to \infty$, $D \to 1$ and the cluster grows as a line across the plane. When $\mu \to 0$ the cluster compacts and $D \to 2$.

In fact, this parameter μ enables a continuum of cluster types to be generated from the linear city form through to the concentric, that is from the one- to the two-dimensional. This was first demonstrated by Niemeyer, Pietronero and Wiesmann [Niem84], but a more detailed and somewhat more technical illustration is provided by the author in a paper complementary to this one [Batt90]. Finally, it is worth noting that there are many variants of these types of model. For example, Meakin [Meak83c] has introduced a family of models in which the occupation of lattice sites is governed by a screening length, which acts as a measure of the influence of a site on surrounding sites. By simple manipulation of the equations embodying this screen, the same continuum of forms as can be generated by the DLA and DBM models results; but as there are several parameters controlling the screen, a more physically diverse set of forms can be generated.

Simulating Cluster Growth Using DLA and DBM

Although the operation of the DLA algorithm is straightforward, the amount of computer time required acts as a constraint on the size of the lattice and the number of particles simulated. Most of the examples in the literature have been developed on comparatively small lattices based on less than a 1000×1000 grid and with cluster sizes less than 50,000 particles. Meakin [Meak86] has produced some mega-DLA simulations with up to one million particles, but such simulations require supercomputers or parallel processors; and in any event, the real systems to which these large simulations might be compared are often only observable at a much coarser level of resolution. Our first work with the DLA model was based on a 500×500 lattice, and the simulation of 10,000 clustered particles was compared with the urban form of the small English town of Taunton (population $\approx 49,000$ in 1981). The fractal dimension D of the DLA simulation which is shown in Plate 5 was estimated from two-point correlation analysis as 1.652, in comparison to the dimension of Taunton which was measured as 1.636. Technical details are given in the paper by Batty, Longley and Fotheringham

[Batt89]; the colour spectrum in Plate 5 shows the order in which the particles were added to the cluster.

This simulation took some 10 hours of CPU time on a Microvax 2 which was dedicated to this work. In contrast, the DBM simulations took much longer because these require iterative solution of the potential equations governing the field each time a particle is added to the cluster. Laplace's equation which describes the potential field is formulated as a discrete approximation to Eq. (12). We define this discrete potential as ϕ_{ij}, where i and j now refer to the row and column numbers of the lattice. Using the method of forward differences, at any site ij on the lattice the potential is an average of the potential at adjacent sites, that is

$$\phi_{ij} = \frac{1}{4}\{\phi_{i+1,j} + \phi_{i-1,j} + \phi_{i,j+1} + \phi_{i,j-1}\} \qquad (16)$$

Equation (16) is solved for ϕ_{ij} iteratively, using a method of overrelaxation which requires about 10 cycles for each additional particle allocated. The time taken to solve the DBM model is much greater than the DLA. For example, the DBM cluster shown in Plate 6 is based on a 300×300 lattice containing 4157 particles, which took some 35 hours of CPU time on the Microvax 2. If the 500×500 lattice containing 10,000 particles shown in Plate 5 had been solved using the DBM model, this would have taken about 12 days of CPU time. Accordingly, to keep the simulation within reasonable bounds a 150×150 lattice was used for the DBM models, and the simulation was terminated once a particle was located some two-thirds the distance from the central seed to the edge of the lattice. In the case of a 150×150 lattice this would be 50 lattice units from the seed, and this system can be simulated in about five hours of CPU time for the DBM model with $\mu = 1$, the pure fractal case.

Properties of the Simulated Clusters

Estimating the Fractal Dimension

The key measure of the extent to which the fractal cluster fills the space available to it is the dimension D, which can vary from 1 to 2 and has a value of 1.71 for the pure fractal case. This dimension is usually estimated from the particle-distance relations given previously in Eqs. (10) and (11). In general, these show that the cumulative number of particles $N(R)$ at a distance R scale as R^D, and the density $\rho(R)$ scales as R^{D-2}. There are also other related functions which can be estimated to give the fractal dimension, for example the branch counts based on $\{dN(R)/dR\}$ proposed by Pietronero, Evertsz and Wiesmann [Piet86].

These relationships can be estimated in two ways. Firstly, $N(R)$ can be counted with respect to the distance R from a single point in the cluster, usually the central seed site. This provides a measure of the population at a distance which has been widely used to describe the form of cities in the mainstream urban studies literature. Density $\rho(R)$ can also be easily calculated, and statistics based on these one-point correlation functions can be estimated using log-log regression on the related variables. Secondly, two-point correlation functions can be calculated if $N(R)$ is counted with respect to every occupied site in the cluster and then averaged. The details of these calculations are given in Batty, Longley and Fotheringham [Batt89], where it is clear that the fractal dimension can be easily derived from the slopes of the associated log-log regressions.

There are two major problems with the use of these statistics. Firstly, there are severe edge effects in the relationships between population and distance. Towards the boundary of the growing cluster, more and more sites which would ultimately be developed have not yet been developed because the cluster is still growing; this has the effect of biasing the fractal dimension downwards. It is therefore necessary to exclude a large area of this growing zone from the calculations. Secondly, there are also volatile effects at short distances from the points around which these various relationships are formed, and these too must be excluded. Because of the inevitable arbitrariness of the cut-off zones chosen, the fractal dimension can vary quite substantially in the same cluster. For the DLA simulation shown in Plate 5, for example, the values of D for the one-point functional relationships ranged from 1.174 to 1.686 for the entire cluster, and from 1.659 to 1.739 for the cluster which excluded the short and long range edge effects. For the two-point relations, D varied from 0.161 to 1.136 for the original cluster, with a much narrower range from 1.640 to 1.677 for the cluster modified to remove the edge effects. The best estimate of the fractal dimension of this cluster was 1.652, based on a near perfect fit ($r^2 = 0.999$) of Eq. (10) to its data.

In Plate 6 the dimensions of the DBM model have been computed in the same way. The unmodified one-point functions produce values of D from 0.994 to 1.317, while their modified form gives a range of dimension from 1.376 to 1.737. For the two-point measures the unmodified range is much wider, from 0.005 to 1.311, while the modified range is from 1.537 to 1.646, with a best estimate of 1.646, giving again an almost perfect fit ($r^2 = 0.999$). One issue which has only been broached in the most casual way is the fact that each simulation will produce a different and unique cluster. Thus, to estimate the best overall dimension, averages over several runs are required. Furthermore, the computation of the two-point correlations takes a substantial amount of CPU time for each cluster (around three hours for the cluster shown in Plate 5). Combined with the volatility of the edge effects and the difficulty of identifying these unambiguously, together with the need to average dimensions over several runs, this has led us to develop a much faster and perhaps more robust method of estimation which we now outline.

Noting that the lattice is square and that the grid size is based on a unitary distance, it is clear that the number of particles at a distance R from the central seed site can be approximated as πR^D. The area associated with this population is also approximately equal to the area of the circle of radius R, that is πR^2, and their ratio gives the density $\rho(R)$, which is

$$\rho(R) \approx \frac{\pi R^D}{\pi R^2} = R^{D-2} \tag{17}$$

For a given value of distance R from the seed site, the value of the fractal dimension at that point, $D(R)$, can be computed from Eq. (17) as

$$D(R) = 2 + \frac{\log \rho(R)}{\log R} \qquad (18)$$

Thus we can compute the variation in $D(R)$ as R increases, and this we refer to, somewhat loosely perhaps, as the 'signature' of the cluster.

Plate 7 shows the baseline fractal cluster associated with the 150 × 150 lattice simulated using the DBM model with $\mu = 1$. The best two-point estimate of the fractal dimension for this cluster is 1.644, with an r^2 of 0.999; but this is subject to all the caveats noted above. In Figure 4 we have graphed the signature of the cluster from Eq. (18), and this is included along with other signatures from the various DBM simulations presented in the next section. The volatile short range effects are quite clearly identifiable from this signature. It is also obvious that the dimension settles down quite quickly to an almost constant value of around 1.7, until it begins to fall gradually as the outer edge of the cluster is reached. The best value of the dimension from the signature is the value of $D(R)$ at its mean. After over 30 runs of the DBM model with $\mu = 1$, this value averages 1.701 ± 0.025.

There are several other properties of these clusters which act to differentiate them. The whole subject of the multifractal nature of such clusters is relevant (see [Stan87a]), but there are some obvious properties relating to average density and nearest neighbours which identify the relative spread and compactness of the form. In Table 1 we have shown these measures computed for the three clusters shown in Plates 5, 6 and 7, for the town of Taunton used in the first simulations developed by the authors [Batt89], and for the town of Cardiff shown later in Plate 10a and presented in a complementary paper [Batt90]. From Table 1 it is clear that the two-point estimates of fractal dimensions are in general less

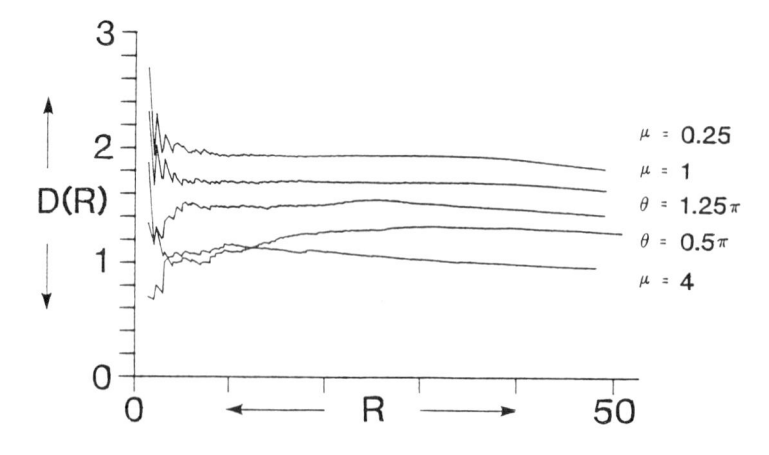

Figure 4. Fractal signatures of the simulated clusters.

Table 1. Comparisons of fractal dimensions and densities.

DLA-DBM System or Real Example	Best Two-point Dimension	Best Signature Dimension	Average Density	Average Nearest Neighbours
DLA 500 × 500 N = 10,000	1.652	nc[†]	0.052	2.394
DBM 300 × 300 N = 4157	1.646	1.708	0.046	4.943
DBM 150 × 150 N = 1856	1.644	1.701	0.082	4.862
TAUNTON 150 × 150	1.636	nc	0.256	4.342
CARDIFF 150 × 150	1.586	1.780	0.278	5.833

[†] nc—the dimension was not computed from the mean of the fractal signature, since the method was developed later [Batt90].

than those computed from the means of the signatures, which in turn are very close to the assumed dimension of such clusters, with $D = 1.71$. It is also clear that the average density of the DLA and DBM clusters are substantially less than those of Cardiff and Taunton. This is largely due to the difficulties over deciding what constitutes urban development and to the coarseness of the maps used to measure this, as well as to the fact that in the urban fields of real cities there are many related but disconnected units of development. These growth models could easily be extended to handle this problem of allowing sites which are physically disconnected but are within the range of influence of the cluster's edge to be accommodated, but this must await further research.

Constraining and Distorting the Growth Process

We have already seen, in the deterministic example of the growing fractal in Figure 2b, that there are many ways in which the units of development could be configured to produce the same fractal dimension of $D = \log(5)/\log(3)$; all that is required is some stable configuration of the five units at each scale being located according to the threefold partitioning of space at the previous level of growth or aggregation. In an analogous way, if the space available for the growing fractal is constrained in some way then the fractal dimension of the same process operating in the same space is likely to be lower. For example, if less than five units of development were only possible in the theoretical fractal in Figure 2b the dimension would be lower. We can broach this problem in our DBM simulations by examining the effect on the dimension of excluding a segment of the circular field around the seed cluster. We can define the space less the segment excluded in terms of the angular variation around the origin. For the DBM simulation in the entire circular field, this uses $\theta = 2\pi$ of arc; as this angular variation available to the growing fractal decreases to $\theta = 0$ the space available collapses to a line, and the fractal is constrained to grow along this line. It is evident that over the scale range used to observe the fractal its dimension D must vary from 1.71 to 1.00, although at finer scales outside this range the fractal can still be displayed as having a dimension of 1.71.

We have demonstrated this effect for two sweeps around the circle, the first based on a space of $\theta = 5\pi/4$ and the second on a space of $\theta = \pi/2$. These are shown in Plates 8a and 8b, respectively, and their fractal signatures are shown in Figure 4. The best estimates of their fractal dimension show the effect of collapsing the available space in this way. Comparing the cases of $\theta = 2\pi$, $\theta = 5\pi/4$ and $\theta = \pi/2$, the fractal dimensions associated with the mean distance in the cluster fall from 1.708 to 1.508 to 1.234. This must be kept in mind when developing DBM models for real city systems. There are many cities which are constrained in this fashion. For example, in Figure 1, although London and Paris are contained in 2π of arc, Tokyo is closer to a space based on $5\pi/4$, while lakeside cities like Chicago and Toronto are close to π degrees of arc. This is not only an issue relating to city systems. There are many physical processes which are so constrained. Even in experiments to simulate such clusters the physical containers within which the experiments take place affect the estimated fractal dimension (see [Nitt85]).

The second type of constraint on the DBM process involves the effect of the scaling parameter μ on the probability distribution governing the candidate sites for development. When this parameter approaches 0 this is akin to agents or developers making little distinction between the peripheral sites, a situation often in accord with the early and rapid growth of small towns. When this parameter increases to a value much greater than 1, developers consider the most attractive sites for development to be at the end of the dendrites which are forcing the growth of the town out into the countryside. This may occur in situations where new innovations in transport technology and infrastructure are being introduced, or it may occur as a result of strong planning policy. We have chosen two simulations from the continuum, with $\mu = 0.25$ and $\mu = 4$; these are illustrated in Plates 9a and 9b, respectively. The forms produced bear out our expectations. The fractal dimensions associated with these two distortions are graphed in the signatures in Figure 4; their best estimates are given as $D = 1.938$ for $\mu = 0.25$, and $D = 1.110$ for $\mu = 4$. An overall idea of the continuums based on the angular physical constraints and the distortion parameter μ can be achieved if Plates 8a, 7and 8b, then Plates 9a, 7, and 9b are examined in sequence.

What these simulations have clearly identified is the difficulty of defining density and development in the context of cities. The best data set for these fractal growth models would be based upon unique and fixed locations for every individual comprising the city, or for cohesive social units such as households. This sort of data is difficult and costly to obtain, but there are also conceptual problems concerning the fixed nature of location. Moreover if the emphasis changes towards physical development, the question of what is development—whether it includes open space around dwellings, whether it includes other types of land use and so on—are all difficult questions which have not been satisfactorily resolved so far. In this way, the value of this approach to city systems enables us to sharpen our research questions rather than to develop predictive simulation models. But in this quest it is important to chart the limits to this style of work. In the following section we thus suggest how fractal growth models might be used to simulate urban form.

Simulating the Growth and Form of Medium-sized Towns: The Example of Cardiff

At some point it is necessary to attempt simulations of real city systems, notwithstanding the obvious limitations of the models discussed so far. One major difficulty with such literal simulation relates to the stochastic nature of the model. Each realisation of the fractal growth process will be different, and any simulation of reality only enables a class of types to be predicted. Insofar as the system being modelled can be seen as typical, then the simulation results will be useful. In cases where specific predictions are required, this type of approach must be extended to enable deterministic simulation. Our initial foray into work in this area involved a simple comparison between the town of Taunton and one realisation of the DLA process on a 500 × 500 lattice. The results of this comparison have already been given in Table 1. Despite the fact that the best dimension estimated from the analysis for Taunton is fairly close to that for the simulation ($D = 1.636$, in comparison with $D = 1.652$), actual comparison of the forms indicates that there are several modifications required before the DLA model can be said to generate a form representative of this town. In fact, the DLA and DBM models would appear to generate the skeleton on which urban growth takes place, thus giving support to the idea that it is the transport network which is perhaps a more suitable system to model using DLA [Beng89].

We now have a way of constraining and distorting the forms produced, by constraining the space within which the fractal or city might grow, and by altering the probability distribution characterising the growing zone so that more compact or sparser clusters can evolve. However, we have not explored the incidence of both effects simultaneously. We know how the parameter μ can enable linear to concentric structures to grow, and how the angular variation θ can reduce the space available for growth. We have not yet examined the impact of these changes together, largely because once one embarks upon such an experiment there are many other valid changes to the DLA-DBM framework which could be suggested. The existence of multiple seed sites, the notion of reversibility, the idea that the cluster may be disconnected but still held together by an invisible glue based on local influence—all these and other possibilities exist.

Nevertheless we have made some preliminary simulations using the town of Cardiff as a basis for comparison. Urban Cardiff has been coded to a 150 × 150 grid based on a grid square of 50 × 50 metres, and the extent of the city is shown in Plate 10a. As in the examples in Figure 1, the urban area is disconnected in its peripheral field. This provides many more units of development than occur in the DLA-DBM simulations, as is clear from Table 1. It is also evident from

Plate 10a that the town is highly constrained physically by its location on the estuary of the River Severn. This suggests that a segment with at least $3\pi/4$ has been removed from its field for potential growth. As in the previous simulations, we have fitted one- and two-point functions to the data with a best estimate of D as 1.586 ($r^2 = 0.997$). But a more reliable guide to this dimension is given by the fractal signature which is plotted in Figure 5. This gives a dimension associated with its mean of 1.786.

Also plotted in Figure 5 is the fractal signature of a baseline simulation using the DBM model in its pure form (that is, with $\mu = 1$). The dimension is somewhat lower than the pure DLA value of 1.71, but this is clearly due to the constraints on the space available; in this sense it is an artifact of the system being simulated. The simulation is shown in Plate 10b, from which it is clear that the form produced is much sparser than the real system. Moreover, the existence of rivers cutting across the field also constrains the extent of urban development. Finally, although the seed initiating growth was planted on the site of the medieval and modern centre, the simulation was unable to develop the port area of the city, which grew almost as a separate centre in the mid- to late nineteenth century. To enable a more realistic simulation, two modifications—one to the data and one to the process—have been made. First, bridging points across the rivers have been input to the data base; second, another seed is planted in the port area after 80 units of development have been generated, thus enabling the simulation to have a fighting chance of generating the docklands area of the town.

Five simulations were attempted; these are shown in Plates 11a to 11e. The differences between the simulations relate to the field parameter μ, which has been set at the values 1.00, 0.75, 0.50, 0.25 and 0.01. The fractal dimensions produced from these simulations range from 1.574 to 1.879 as μ changes from 1.00 to 0.01. The dimensions produced by the nonconstrained theoretical model vary from 1.701 to 1.971 for this range, and it is clear that the physical constraints do reduce the value of this dimension. In another sense, Plates 11a to 11e show forms which are far too tidy to be representative of cities. The possibility does exist, however, of introducing additional seed sites based on historic urban cores and villages which get swallowed up in the urban expansion. It is also possible to lay out the simulation on a nonhomogeneous plane which reflects the underlying topography and geology of the region. There is also the possibility that a degree of reversibility could be built into the diffusion process. All these extensions would increase the realism of the simulation. As such they represent important directions for future research.

Conclusions

Without computers and computer graphics the study of fractals would be quite impossible. So many systems of interest require on the one hand extraordinary amounts of repetitive calculation, and on the other hand powerful methods for visualising their output, that any scientific analysis is inseparable from its means of computation. In the study of fractals this may seem self evident, although it does have important implications for the way we research such systems. Such systems seem to have a richness which makes it more important than is usual to resist extensive computer experimentation. It is a simple matter to add more and more complexity to the models which are being developed here, but it is essential to maintain these models in their most parsimonious form if useful and general insights are to be generated. For example, in the DLA model it is only just becoming apparent after nearly 10 years of research that the fractal dimension of 1.71 of the cluster produced may ultimately be scale or at least size dependent, thus casting considerable doubt on the claims for its universality which heralded earlier research. Thus, the need to simulate ever bigger systems is an important priority. In this, the emergence of a new generation of mini-supercomputers and parallel processors is likely to play an important part. Many of the research directions alluded to in this paper make access to such computers essential.

There are, however, a number of research directions related to the growth process which should be incorporated in new models applied in this domain. Reversible systems need to be researched as soon as possible to see how sensitive the simple irreversible growth models are to such change. The extent to which cities might compete with one another based on growth from different seeds is an area which would produce useful insights, but very careful control is required if such experiments are to yield useful results. It would not be useful, for example in the case of Cardiff, to plant seeds all over its region, representing all the historic cores which have been incorporated into the growth of the town. First it is necessary to experiment with hypothetical structures in which growth regimes are under total control.

The possibility also exists that these types of physical simulation might act as shells within which more conventional and acceptable socioeconomic urban models might be run. One of the problems in contemporary urban and transport modelling is the difficulty of linking the behaviours of activities and groups of like

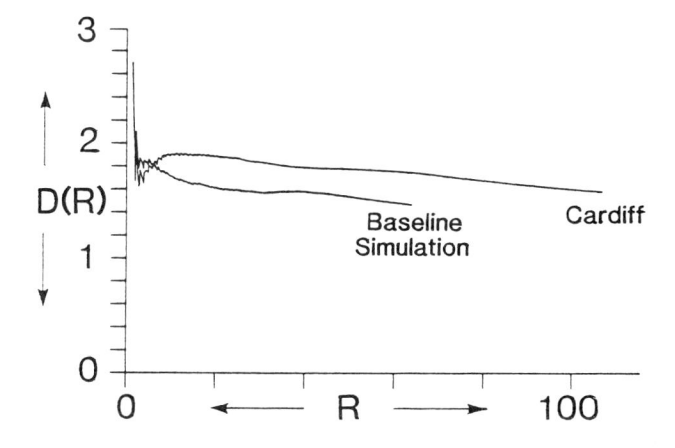

Figure 5. Fractal signatures of Cardiff and its baseline simulation.

actors to their physical impact on the form of the city. The translation from activities to land uses to sites, thence to physical structures, represents an area of research which is poorly understood. It is thus essential to attempt to forge strong and consistent links between physical form and social and economic processes.

In the application of fractal ideas to any domain there is always something which is relevant to research into fractals *per se*. This domain is no different from those discussed elsewhere in this book. The very definition of the idea of a fractal comes under close scrutiny in any application. For example, the notion that a fractal is independent of scale, at least over a significant range of scale, can be clearly identified with respect to built form. Although one criterion for defining a fractal is the stability of its self-similarity across a range of scales, whenever a fractal is measured there is a base scale on which all other relevant scales must be predicated. This is clearly seen in Figure 2 in this paper, where the fractal can be seen at the base scale over a total of four orders of scale. If these orders are extended up or down scale they cannot be detected with the given base scale, but this does not mean that they do not exist. In constraining the available space in which a fractal might grow, the dimension of the fractal changes but only with respect to the base and its associated finer scales. In this sense a fractal depends upon a limited range of scales within the measuring instrumentation available, and changes to its dimension by constraining these scales only apply to the base scale from which such measurement takes place. In this practical sense, then, a fractal is always scale dependent.

Finally, it is worth emphasising the value of the approach taken here. Our quest is very definitely not to produce workable predictive spatial models of cities. Such models already exist within certain limits, as we have alluded to in this paper. But the fractal approach to cities is of enormous value in sharpening critical research questions and focussing upon difficult but important issues. This approach has raised the critical but poorly understood problem concerning 'what is density, and how can it be measured?' as well as the central issues regarding the appropriate scale for its definition. It is somewhat remarkable that there has been so much work on measuring population density in cities with so little concern for what is being measured and modelled. Fractals have a clear role to play here. As Marc Kac [Kac69] so cogently observed: 'The main role of models is not so much to explain or predict—although these are the main functions of science—as to polarize thinking and to pose sharp questions'. It is in this spirit that this work should be interpreted and future work in this field should be conducted.

References

[Arli85]
Arlinghaus, S.L., Fractals take a central place, *Geografiska Annaler*, Vol. 67B, pp. 83–88, 1985.

[Batt86]
Batty, M., and Longley, P.A., The fractal simulation of urban structure, *Environment and Planning A*, Vol. 18, pp. 1143–1179, 1986.

[Batt88]
Batty, M., and Longley, P.A., The morphology of urban land use, *Environment and Planning B*, Vol. 15, pp. 461–488, 1988.

[Batt89]
Batty, M., Longley, P.A., and Fotheringham, A.S., Urban growth and form: Scaling, fractal geometry, and diffusion-limited aggregation, *Environment and Planning A*, Vol. 21, pp. 1447–1472, 1989.

[Batt90]
Batty, M., Generating urban forms from diffusive growth, submitted to *Environment and Planning A*, Vol.23, forthcoming 1991.

[Beng89]
Benguigui, L., and Daoud, M., Is the suburban railway system a fractal?, unpublished paper, Laboratorie Leon Brillouin, CEN Saclay, France, 1989.

[Butt85]
Buttenfield, B., Treatment of the cartographic line, *Cartographica*, Vol. 22, pp. 1–26, 1985.

[Chap62]
Chapin, F.S., and Weiss, S.F., Eds., *Urban Growth Dynamics: In a Regional Cluster of Cities*, New York: John Wiley and Sons, 1962.

[Chri66]
Christaller, W., *The Central Places of Southern Germany*, translated by C.W. Baskin, Englewood Cliffs, NJ: Prentice-Hall, 1966.

[Doxi68]
Doxiadis, C., *Ekistics: The Science of Human Settlement*, London: Hutchinson, 1968.

[Dutt73]
Dutton, G., Criteria of growth in urban systems, *Ekistics*, Vol. 215, pp. 298–306, 1973.

[Fede88]
Feder, J., *Fractals*, New York: Plenum Press, 1988.

[Good87]
Goodchild, M.F., and Mark, D.M., The fractal nature of geographic phenomena, *Annals of the Association of American Geographers*, Vol. 77, pp. 265–278, 1987.

[Kac69]
Kac, M., Some mathematical models in science, *Science*, Vol. 166, pp. 695–699, 1969.

[Mand67]
Mandelbrot, B.B., How long is the coast of Britain? Statistical self-similarity and fractional dimension, *Science*, Vol. 156, pp. 636–638, 1967.

[Mand82]
Mandelbrot, B.B., *The Fractal Geometry of Nature*, New York: W.H. Freeman, 1982.

[Meak83a]
Meakin, P., Diffusion-controlled cluster formation in two, three and four dimensions, *Phys. Rev. A*, Vol. 27, pp. 604–607, 1983.

[Meak83b]
Meakin, P., Diffusion-controlled cluster formation in 2–6 dimensional space, *Phys. Rev. A*, Vol. 27, pp. 1495–1507, 1983.

[Meak83c]
Meakin, P., Cluster-growth processes on a two-dimensional lattice, *Phys. Rev. B*, Vol. 28, pp. 6718–6732, 1983.

[Meak86]
Meakin, P., Computer Simulation of Growth and Aggregation Processes, in *On Growth and Form: Fractal and Non-Fractal Patterns in Physics*, Stanley, H.E., and Ostrowsky, N., Eds., Boston, MA: Martinus Nijhoff, pp. 111–135, 1986.

[Mull86]
Muller, J-C., Fractal dimension and inconsistencies in cartographic line representations, *Cartographic Jour.*, Vol. 23, pp. 123–130, 1986.

[Mull87]
Muller, J-C., Fractal and automated line generalization, *Cartographic Jour.*, Vol. 24, pp. 27–34, 1987.

[Muth83]
Muthukumar, M., Mean field theory for diffusion-limited cluster formation, *Phys. Rev. Lett.*, Vol. 52, pp. 839–842, 1983.

[Niem84]
Niemeyer, L., Pietronero, L., and Wiesmann, H.J., Fractal dimension of dielectric breakdown, *Phys. Rev. Lett.*, Vol. 52, pp. 1033–1036, 1984.

[Nitt85]
Nittman, J., Daccord, G., and Stanley, H.E., Fractal growth of viscous fingers: Quantitative characterization of a fluid instability phenomena, *Nature (London)*, Vol. 314, pp. 141–144, 1985.

[Nyst66]
Nysteun, J.D., Effects of boundary shape and the concept of local convexity, Discussion Paper 10, Michigan Inter-University Community of Mathematical Geographers, Department of Geography, University of Michigan, Ann Arbor, MI, 1966.

[Penc94]
Penck, A., *Morphologie der Erdoberflache*, Stuttgart, Germany, 1894 (quoted in [Perk58a]).

[Perk58a]
Perkal, J., On the length of empirical curves, *Zastosowania Matematyki*, Vol. 3, pp. 257–286 (translated by W. Jackowski and reprinted in Discussion Paper 10, Michigan Inter-University Community of Mathematical Geographers, Department of Geography, University of Michigan, Ann Arbor, MI), 1958.

[Perk58b]
Perkal, J., An attempt at objective generalization, *Geodezja i Kartografia*, Vol. 7, pp. 130–142 (translated by W. Jackowski and reprinted in Discussion Paper 10, Michigan Inter-University Community of Mathematical Geographers, Department of Geography, University of Michigan, Ann Arbor, MI), 1958.

[Piet86]
Pietronero, L., Evertsz, C., and Wiesmann, H.J., Scaling Properties of Growing Zone and Capacity of Laplacian Fractals, in *Fractals in Physics*, Pietronero, L. and Tosatti, E., Eds., Amsterdam: North Holland, pp. 159–163, 1986.

[Rich61]
Richardson, L.F., The problem of contiguity: An appendix to 'The statistics of deadly quarrels', *General Systems Yearbook*, Vol. 6, pp. 139–187, 1961.

[Sand87]
Sander, L.M., Fractal growth, *Sci. Amer.*, Vol. 256, pp. 82–88, 1987.

[Sher86]
Sherwood, J., and Nittman, J., Gradient governed growth: The effect of viscosity ratio on stochastic simulations of the Saffman-Taylor instability, *Jour. de Physique*, Vol. 47, pp. 15–22, 1986.

[Stan87a]
Stanley, H.E., Multifractals, in *Time-Dependent Effects in Disordered Materials*, Pynn, R., and Riste, T., Eds., New York: Plenum Press, pp.145–161, 1987.

[Stan87b]
Stanley, H.E., Role of Fluctuations in Fluid Mechanics and Dendritic Solidification, in *Time-Dependent Effects in Disordered Materials*, Pynn, R., and Riste, T., Eds., New York: Plenum Press, pp.19–43, 1987.

[Stan88]
Stanley, H.E., and Meakin, P., Multifractal phenomena in physics and chemistry, *Nature (London)*, Vol. 335, pp. 405–409, 1988.

[Stei54]
Steinhaus, H., Length, shape and area, *Colloquium Mathematicum*, Vol. 3, pp. 1–13, 1954.

[Stei60]
Steinhaus, H., *Mathematical Snapshots*, Oxford, UK: Oxford Univ. Press, 1960.

[Stev74]
Stevens, P.S., *Patterns in Nature*, New York: Little, Brown, 1974.

[Stew58]
Stewart, J.Q., and Warntz, W., Physics of population distribution, *Jour. Reg. Sci.*, Vol. 1, pp. 99–123, 1958.

[Vics89]
Vicsek, T., *Fractal Growth Phenomena*, Singapore: World Scientific Co. Pte., 1989.

[Witt81]
Witten, T.A., and Sander, L.M., Diffusion-limited aggregation: A kinetic critical phenomenon, *Phys. Rev. Lett.*, Vol. 47, pp. 1400–1403, 1981.

[Witt83]
Witten, T.A., and Sander, L.M., Diffusion-limited aggregation, *Phys. Rev. B*, Vol. 27, pp. 5686–5697, 1983.

[Wold67]
Woldenberg, M.J., and Berry, B.J.L., Rivers and central places: Analogous systems?, *Jour. Reg. Sci.*, Vol. 7, pp. 129–139, 1967.

[Wold73]
Woldenberg, M.J., An allometric analysis of urban land use in the United States, *Ekistics*, Vol. 215, pp. 282–290, 1973.

[Zipf49]
Zipf, G.K., *Human Behavior and the Principle of Least Effort*, Reading, MA: Addison-Wesley, 1949.

Modelling Growth Forms of Sponges with Fractal Techniques

J.A. Kaandorp

Abstract

This paper describes how fractal modelling techniques can be applied for generating morphological models of biological objects. For a case study the growth forms of the sponge species Haliclona oculata *(Porifera) are used. From simple ramifying fractals, more developed models are created by adding more rules to the generation process, with which growth of a sponge in 2D can be simulated. These models, in a more evolved form, might be useful for ecological and taxonomic research. The models are generated with an interactive fractal system in which geometric constructions are used for the specification and calculation of the (fractal) objects.*

Introduction

Fractals form a powerful tool for generating 'objects' which show a high resemblance to biological objects. The fractal dimension can be demonstrated for many biological objects, for example for the surface of the brain [Mand83], coral reefs [Brad83], proteins [Lewi85], vegetation [Mors85].

Fractals are defined by B.B. Mandelbrot in his book *The Fractal Geometry of Nature* [Mand83] as geometric objects whose Hausdorff-Besicovitch dimension exceeds its topological dimension; in most cases these objects have a fractional Hausdorff-Besicovitch dimension. In Barnsley [Barn88] several numbers are discussed which can be associated with fractal objects; they are referred to as fractal dimensions. An important feature of fractals is that they are (statistically) self-similar on each scale.

Many methods have been described in computer graphics for generating objects with a resemblance to biological objects. In Kawaguchi [Kawa82] a method is described for creating images of corals and shells. Impressive examples of images of biological objects generated with computer graphics can be found in Bloomenthal [Bloo85], Oppenheimer [Oppe86] and Barnsley [Barn88]. In the last two references, fractal techniques were applied for generating the images. Although the objects shown in the papers cited above sometimes have an amazing resemblance to real objects, the procedure for their generation is not based on a model of a biological process. The plant images are not generated, for example, by modelling the growth of meristems but by artificial constructions.

A well-known technique which has been developed for describing biological objects is known as L-systems. The original description of L-systems can be found in Lindenmayer [Lind68]. Studies in which L-systems were applied are Hogeweg and Hesper [Hoge74]; also the images of plants and trees in Aono and Kunii [Aono84] were obtained by using L-systems. In Prusinkiewicz et al. [Prus88] and de Reffye et al. [deRe88], developmental models based on L-systems are presented. With L-systems it is possible to make growth models of filamentous organisms and to describe branching patterns of filamentous organisms. Growth models of organisms represented in two dimensions (for example, growth of meristems) can less conveniently be described with L-systems.

An example of a study where fractals are applied for growth models of biological objects can be found in Meakin [Meak86]. In this paper a method related to cellular automata is used. Growth models like this, known in the literature as 'Diffusion Limited Aggregation' (DLA) models [Witt81], are applied widely in physics (see [Stan87] and [Fede88]). With DLA models the objects grow in a lattice. For many biological objects, however, it is hard to describe growth in a lattice.

Basically, growth processes can be described as an iteration process (Figure 1), where the last stage of a growing object serves as input for the next stage. In the growth process, after each iteration new material is added to the object.

The same iteration process can be used for the generation of fractal objects. The relation between the output and the input in an iteration process may be linear or nonlinear. Examples of fractal objects generated by an iteration process with a nonlinear relation are the famous complex quadratic mappings. The result of the iteration process is not necessarily an object with a fractal dimension; the process may also yield nonfractal figures or result in points at infinity or in single points of attraction.

The technique which is applied in this paper for generating objects uses geometric constructions. The iteration process starts with an initial polygon (in 2D

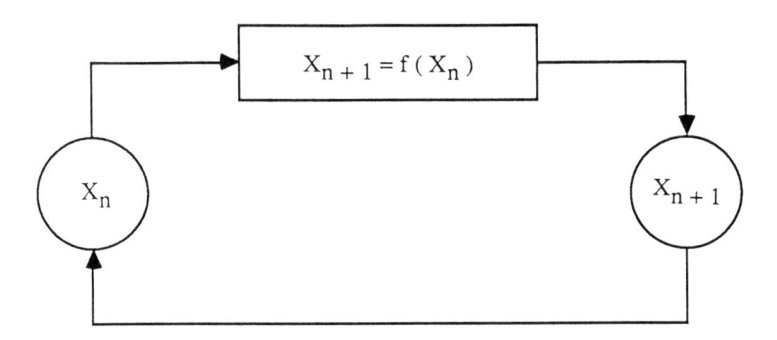

Figure 1. Diagram of an iteration process where the output of one iteration is used as input for the next one.

constructions), 'the initiator', X_0. In 3D constructions this initiator might, for example, be a volume. During each iteration certain sides of the initiator, which are indicated as fertile sides, or another stage in the construction (the fractal-approximant, X_n) are replaced by the replacing polygon/volume: 'the generator'. The function $f(X_n)$ is embodied by a geometric production rule. This rule consists of a series of transformations that can be written as a linear mapping. An example of such a geometric production rule is shown in Figure 2. The resulting ramifying fractal is shown in Plate 12. An advantage of this technique is that geometric production rules can be designed interactively in an easy way. The production rules already contain the geometric information; 2D geometric production rules can relatively easily be extended to 3D for spatial objects. This makes the last method very suitable for morphological growth models.

In the first part of this paper some examples of ramifying fractals will be shown, that are all derived from the ramifying fractal shown in Plate 12. The growth process of these ramifying fractals has no biological relevance, but several aspects of growth processes can easily be demonstrated with this class of fractals. In the biological literature comparable ramifying fractals are used to simulate branching patterns (see [Bell86]).

In the second part of this paper it is shown how a subset of the rules applied in the first part can be used for simulating the growth process of organisms such as sponges (Porifera) and corals (Scleractinia). The biological interpretation of the rules applied in these models is discussed briefly; the details will be discussed in another paper [Kaan, in prep.]. One important reason for the choice of sponges and corals as subjects of a case study is that these organisms exhibit a relatively simple growth process, which makes it comparatively easy to design geometric production rules to simulate growth processes. In the section about the sponge models, one sponge species, the species *Haliclona oculata*, is used in particular as an example.

The models of ramifying fractals and sponges shown in this paper were generated with an interactive *fractal system*. The details of this fractal system are

initiator	generator

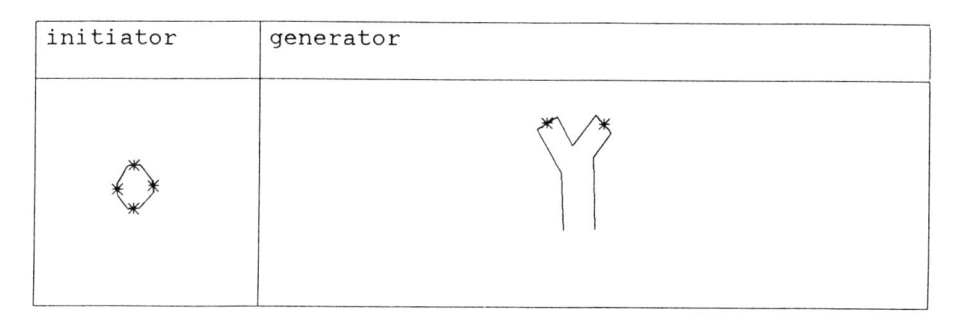

Figure 2. Geometric production rule for a ramifying fractal. Fertile sides (sides of the preceding fractal approximant, which will be replaced in next iteration steps) are marked with asterisks. The resulting fractal is shown in Plate 12.

74   J.A. Kaandorp

discussed by Kaandorp [Kaan87]. The fractal system was implemented using standard graphics systems such as GKS (see Hopgood et al. [Hopg83]), PHIGS (see Shuey et al. [Shue86]) in the C programming language. In this system geometric production rules are used for the specification, design and calculation of fractal objects.

The Ramifying Fractals

The first extension is the introduction of functions which process the original generator. The original generator is represented internally in the fractal system as a sequence of transformations, the *generator_function*. The processing functions will be indicated as 'generator processing functions' (*generator_proc*). A generator processing function which will generate a ramiformous fractal with an irregular appearance uses the original generator as an argument and delivers a new generator on which a rotation is performed, the angle of rotation being chosen randomly between two limits (see Figure 3). The result of this production rule is shown in Plate 13.

In the case of the irregular ramifying fractal, only the original generator is used as an argument in the generator processing function. When an irregular ramifying fractal is generated that is growing in a certain prevailing direction, some additional local information (indicated as *local_inf*) is necessary. In the present case this local information is the angle of a fertile element with the prevailing direction of growth. The probability that the object is branching towards the prevailing direction is maximal when this angle is $\pi/2$ and decreases towards a minimum value at an angle of $3\pi/2$. An example of an irregular ramifying fractal growing in a prevailing direction (in this case the right upper corner) is shown in Plate 14.

initiator	generator
◇	Y Y

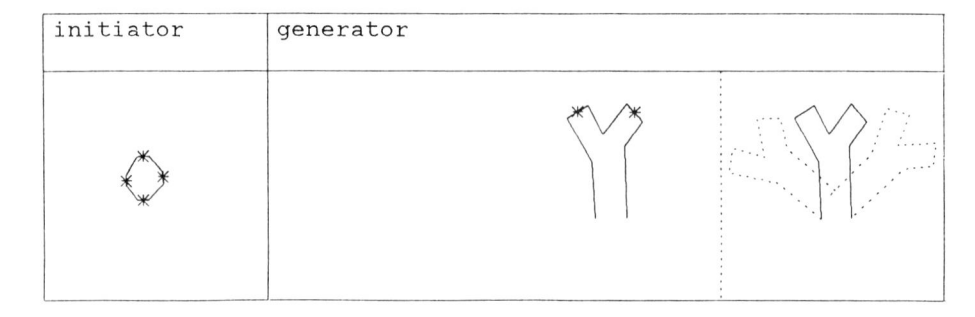

Figure 3. Geometric production rule for an irregular ramifying fractal. The original generator is processed by a generator processing function, which allows random movements of the generator between two limits. The generator processing function is described in the right part of the generator component.

The second extension is the introduction of functions which postprocess the fractal approximant. These functions will be indicated as 'postprocessing functions' and are referred to in this paper with the prefix *post_*... An example is a function which removes the intersecting branches from the fractal approximant. The result, a nonintersecting, ramifying fractal, is shown in Plate 15. The two extensions are represented in a new diagram of the iteration process shown in Figure 4, where a chain of postprocessing functions ($post\_proc_0, post\_proc_1, \cdots, post\_proc_n$) is shown.

The chain of postprocessing functions may contain all kinds of rules for the growing object. Without these postprocessing functions the elements of a fractal approximant cannot interact with each other; they make it possible to introduce restrictions—for example, an element is not allowed to intersect with another element, or elements should remain at a certain distance from each other.

In Plate 15 growth stops as soon as a branch intersects a part of the object. In Figure 5 the iteration process of Figure 4 is extended with an additional cycle. In this new iteration process, a fertile site produces a new branch that is tested for intersections with the object. When an intersection is found a new branch is generated, until a branch is found, by trial and error, that does not intersect, or until the number of attempts exceeds a certain maximum number (*nretry*). In the last case the branch is removed from the object, and the tested fertile site becomes nonfertile. In Plate 16 the result of this new addition is shown. In this figure each fertile site produces a maximum number of new branches (in this picture *nretry* = 10 was used). The result is that branches try to avoid each other and that the number of created branches is higher than in Plate 15.

For the generation of Plate 17 a new postprocessing rule is inserted in the chain of postprocessing rules, stating that the fractal approximant is not allowed to intersect a certain object. In Plate 17 the object is a nearly closed box. The aperture of this box is found, by trial and error, by the ramifying fractal.

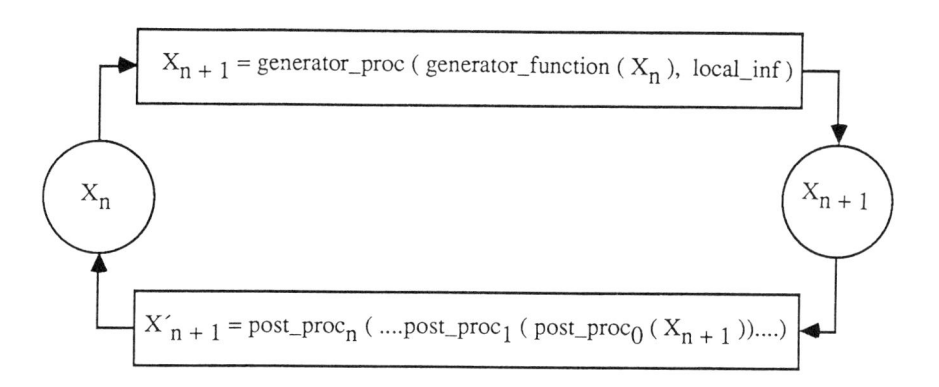

Figure 4. Diagram of the iteration process, in which two extensions are introduced in the original iteration process (Figure 1): the generator processing function (*generator_proc*), a chain of postprocessing functions (*post_proc$_x$*).

76   J.A. Kaandorp

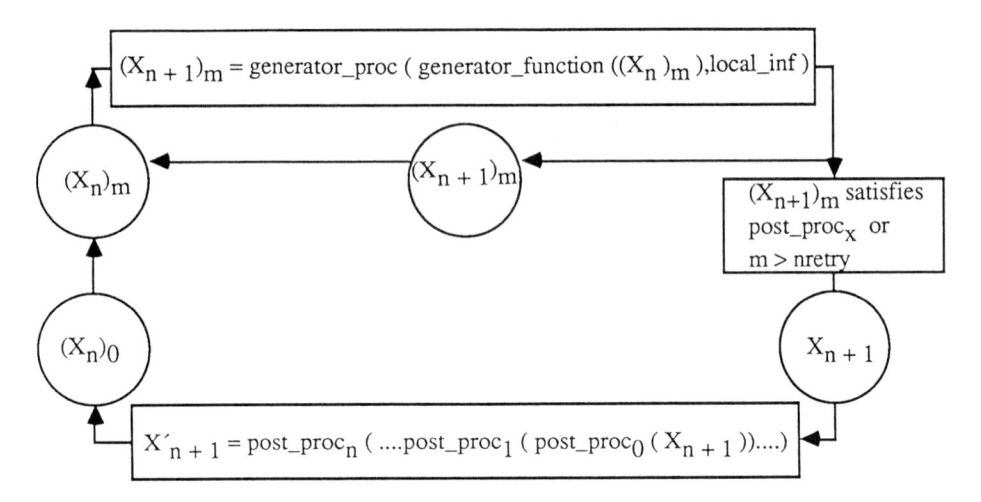

$(X_{n+1})_m$ = generator_proc (generator_function $((X_n)_m)$, local_inf)

$(X_n)_m$

$(X_{n+1})_m$

$(X_{n+1})_m$ satisfies post_proc$_x$ or $m >$ nretry

$(X_n)_0$

X_{n+1}

X'_{n+1} = post_proc$_n$ (....post_proc$_1$ (post_proc$_0$ (X_{n+1}))....)

Figure 5. Diagram of the iteration process, where the iteration process (Figure 4) is extended with an additional cycle.

The Sponge Models

Growth Forms and Internal Architecture of *Haliclona Oculata*

The sponge species used in this case study, *Haliclona oculata* (see also [Hart58] and [deWe86]), shows a wide range of growth forms (see Figure 6). The overall body shape of *Haliclona oculata* tends to be more or less flattened, although forms which are branching in all directions often appear as well. Growth form A is typical for sheltered sites. The sponge is almost dichotomously branching and forming thin tubes. Growth form B is typical for more exposed sites. It shows plate-like growth forms. Forms A and B are two extremes. The exhibited forms are young examples of this species; between both forms all kinds of intermediates can be found. Deviations from this general trend easily arise as a result of damage to the sponge in the course of time. More irregular forms are found among sponges of several years' age. Probably tissue material was removed by abrasion, and irregularities occurred when new material was added.

The phenomenon that a single species exhibits a wide range of growth forms (phenotypes), often caused by environmental differences, is a common problem in the taxonomy of sponges and corals. In order to describe growth forms in terms less subjective than 'thin tubes', 'plate-like', 'irregular', it is necessary to develop methods for describing growth forms with mathematical models. One way of avoiding subjective morphological terms could be the calculation of parameters as the fractal dimension (see [Kell87], [Pent84] and [Barn88]). The significance of the fractal dimension is limited; as a given value it might be valid for a large class of objects. Growth models, especially models which are based on the actual growth process, could be very useful in obtaining a better understanding of these

Modelling Growth Forms of Sponges With Fractal Techniques   77

(a)   (b)

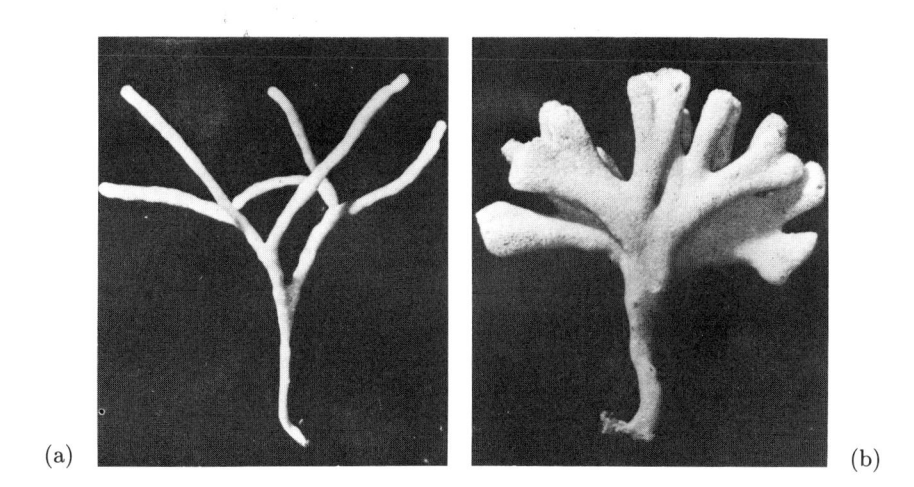

Figure 6. Two extreme growth forms of the sponge species *Haliclona oculata* (natural size). Form A is typical for sheltered sites, form B for more exposed sites.

forms. These models in a more evolved form might be useful for ecological and taxonomic research. Examples of studies in which computer models were used to describe the different growth forms of the coral species *Montastrea annularis* (Scleractinia) are Graus and Macintyre [Grau82] and Graus [Grau77].

The growth model, which will be discussed below, is based on the skeleton architecture and the aquiferous system. The skeleton architecture and the aquiferous system represent the basal patterns in growth forms. Together with the environmental factors they are the main parameters in causal explanations of the growth forms of sponges.

The skeleton of *Haliclona oculata* consists of skeleton elements (the spicules) which are connected with a special protein (spongin) and are consolidated in a 3D mesh. The type of skeleton architecture is known as 'radiate accretive' (terminology after Wiedenmayer [Wied77]). The skeleton can be visualized by drying and sectioning the sponge. The effect of drying is that all material except the spicules more or less shrinks away. An overall view of the skeleton, as seen through a stereo microscope, is shown in Figure 7. In the radiate accretive architecture it is possible to discriminate ascending (longitudinal) fibres and tangential fibres. The radiate accretive architecture is developing in a growth process, in which a new layer of material is added at a tip of a branch or column. The tangential tracts may correspond to surfaces of earlier growth stages (the growth lines). The 3D mesh of spicules in a tip of this sponge shows a radial symmetry; a longitudinal section (parallel to the axis of the tip) will always show about the same. The tips might only be somewhat flattened.

The aquiferous system of *Haliclona oculata*, the system with which the sponge is pumping water together with suspended material through the tissue, is only poorly developed. A consequence is that this species never can develop massive globular forms, with tissue far away from the environment.

78   J.A. Kaandorp

Figure 7. Simplified version of a longitudinal section through a branching tip of *Haliclona oculata* showing a radiate accretive architecture. The growth lines and longitudinal fibres are represented by lines.

Development of Growth Rules for Sponge Models

In this section the development of growth rules is discussed. The relation between these rules and the real biological objects is indicated briefly.

The basic construction: the *generator_function*

The simplest construction is obtained with the geometric construction shown in Figure 8. In this construction the longitudinal and tangential fibres are represented by lines. The initiator consists of a number of fertile tangential sides

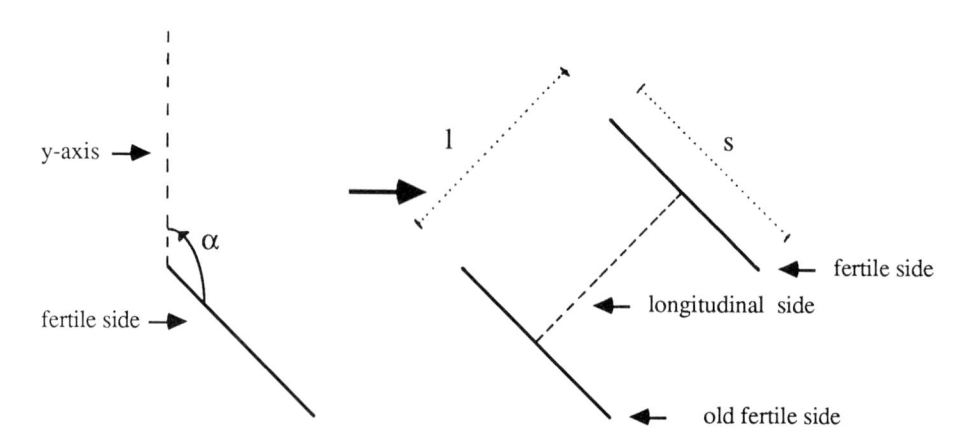

Figure 8. Geometric construction of the sponge model applied in the *generator_function*. The length l of the longitudinal side is determined using a growth function (for example Eq. 1); s is the length of a tangential side.

situated in a hemicircle. The generator replaces each fertile side (length s) by a nonfertile longitudinal side and a fertile tangential side. This geometric construction is indicated in Figure 4 as the *generator_function*. The length of the longitudinal side l is determined by a generator processing function *generator_proc*. This function uses the original *generator_function* and *local_inf* as arguments. The latter is the angle α between the original tangential side and the vertical y-axis. In Eq. (1)

$$f(\alpha) = \sin(\alpha) \qquad \pi/2 \leq \alpha \leq \pi$$
$$l = s \cdot f(\alpha) \qquad (1)$$

l is calculated.

For erect growth forms of filter-feeding organisms like *Haliclona oculata*, it can be assumed that the tip of the sponge has the highest access to suspended material in the water column. Turbulence and other water movements will be strongest at the tip (see [Jack79]); the resulting high amount of food supply of the sponge at this site may give rise to the highest growth velocity. The growth velocity depends upon the angle between the tangent to the surface and the axis of a column-shaped sponge tip.

The part of the *generator_proc* in which l is calculated will be indicated as the 'growth function'. The result of this construction appears at the bottom of the diagram in Figure 9 (form A), in which an overview is presented of the development of the growth models of *Haliclona oculata*.

Modelling the coherence of the skeleton

In order to obtain continuous growth lines in the model it is necessary to postprocess the fractal approximant, and to introduce a new rule ensuring that neighbouring tangential elements be situated on a continuous curve (*post_cont*). In reality (see Figure 7) tangential and longitudinal elements are arranged in continuous lines, caused by the fact that the spicule bundles form one connected mesh.

A second postprocessing rule which is necessary to obtain continuous growth lines is the addition of new tangential elements when there is enough space between two neighbouring elements (*post_adding*). The biological interpretation of the rule *post_adding* is that without adding new tangential elements gaps would appear in the 3D mesh, and the skeleton would disintegrate. By applying both postprocessing rules, the object in Figure 9 (form B) with a coherent skeleton and continuous growth lines can be obtained from the object in Figure 9 (form A).

Introduction of the smallest skeleton element in the model

To generate columns which do not accumulate material at the lateral sides (Figure 9, form B) it is necessary to introduce a new rule in the generator processing function: when l drops below a certain level *inhibition_level*, growth will

80   J.A. Kaandorp

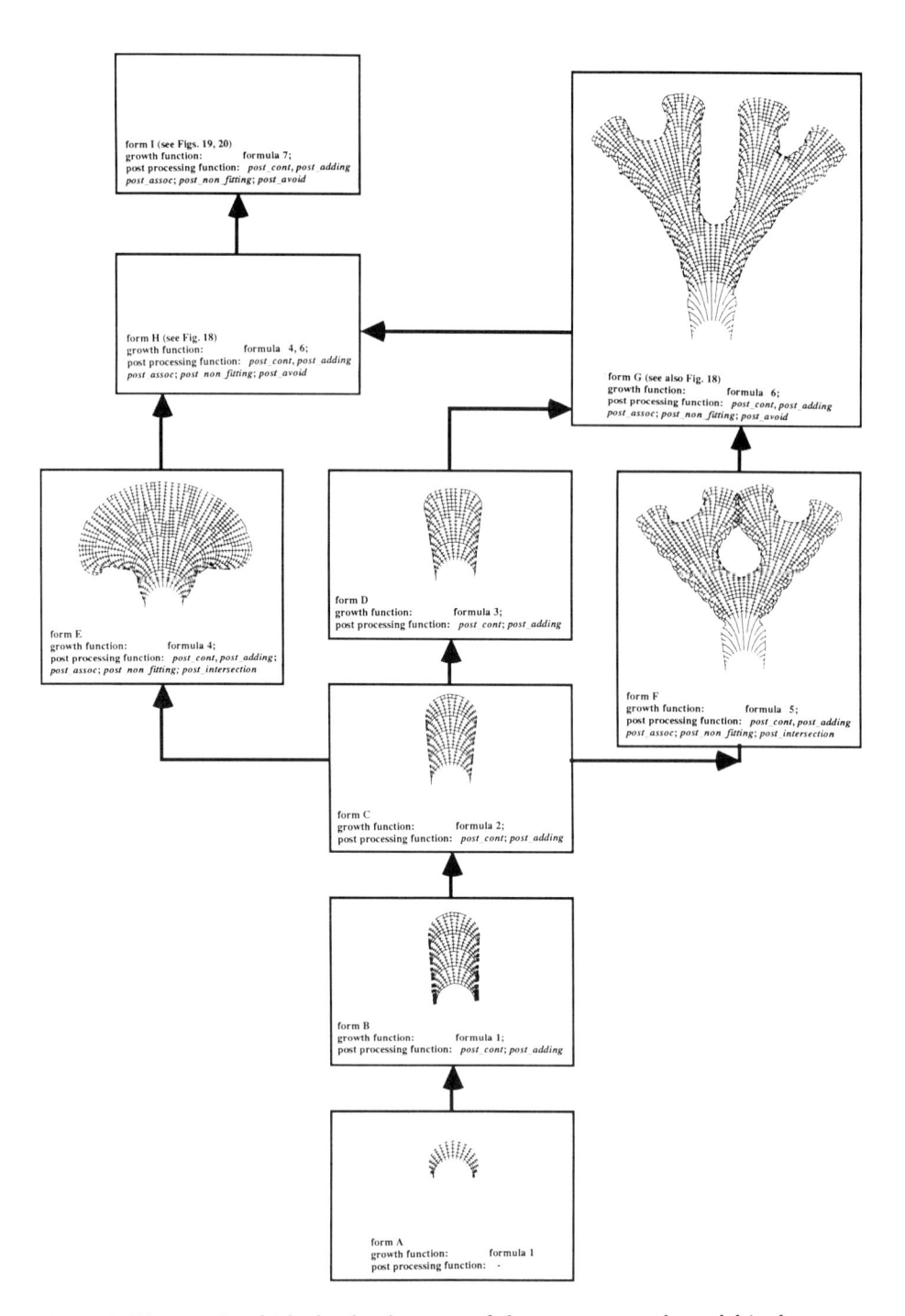

Figure 9. Diagram in which the development of the sponge growth model is shown.

stop. Equation (2)

$$f(\alpha) = \sin(\alpha) \qquad \pi/2 \leq \alpha \leq \pi$$

$$l = \begin{cases} s \cdot f(\alpha) & f(\alpha) > inhibition\_level \\ 0.0 & f(\alpha) \leq inhibition\_level \end{cases} \qquad (2)$$

includes this stopping rule. The result is shown in Figure 9 (form C). When growth of a fertile side is inhibited the status of this fertile side is changed into nonactive. This type of side will be indicated in this paper as nonactive fertile. The growing object is limited by active and nonactive fertile sides; only active sides will participate in generating new sides during subsequent growth stages. The biological interpretation of the threshold value $inhib\_tol$ in this formula is the length of one skeleton element, a spicula.

Modelling the 'widening effect' of a thin branching sponge

The column shown in Figure 9 (form C) can be flattened by using a generator processing function in which an area of equal value appears in the growth function, Eq. (3)

$$f(\alpha) = \begin{cases} 1.0 & \pi/2 \leq \alpha \leq (\pi/2 + \pi/n) \\ \sin((\pi/2)/(\pi/2 - \pi/n) \cdot (\pi - \alpha)) & (\pi/2 + \pi/n) < \alpha \leq \pi \end{cases}$$

$$l = \begin{cases} s \cdot f(\alpha) & f(\alpha) > inhibition\_level \\ 0.0 & f(\alpha) \leq inhibition\_level \end{cases}$$

$$n > 2 \qquad (3)$$

instead of one maximum. The shape can be more or less flattened by choosing different values for n in Eq. (3) (see Figure 9, form D). In a longitudinal section of a thin-branching sponge tip (see Figure 7) it can be observed that the tip widens before it splits into two branches. Without widening, the tip would split into smaller branches (see Plate 12) and attain a certain limit.

Formation of new growth axes

So far growth rules were formulated for columnar forms. These forms have in common one growth axis, the y-axis. In order to create branching forms it is necessary to formulate a new rule, allowing the generation of new growth axes. New growth axes might arise on the surface of the growing object on sites where a (local) maximum in growth velocity occurs. In Figure 10 a longitudinal section of a growing object is shown, in which two local maxima develop. The first seven layers in the object are fertile tangential elements, associated with a single growth axis A. In the eighth layer two local maxima and one local minimum develop. In the eighth and following layers, fertile elements are associated with new growth axes B (the fertile elements to the left of the local minimum) and

C (the elements to the right of the local minimum). The association of fertile elements takes place in a new postprocessing function named *post_assoc*. The angle α between a growth axis and a fertile tangential element is calculated in the generator processing function. Equation (2) is used as the growth function.

Disturbance of the growth process—formation of plates

The growth function described in Eq. (2) will never generate more than one maximum. Maxima can arise when the growth process is disturbed by external influences. A simple example is the superposition of 'noise' on the final length l of the longitudinal side. In Eq. (4) a function g is introduced

$$f(\alpha) = \sin(\alpha) \qquad \pi/2 \leq \alpha \leq \pi$$

$$g(lowest\_value, 1.0) = random\_function(lowest\_value, 1.0)$$
$$\text{for} \quad lowest\_value \leq g() \leq 1.0$$
$$lowest\_value > inhibition\_level$$

$$l = \begin{cases} s \cdot f(\alpha) \cdot g() & f(\alpha) \cdot g() > inhibition\_level \\ 0.0 & f(\alpha) \cdot g() \leq inhibition\_level \end{cases} \qquad (4)$$

which returns random values between two limits, *lowest_value* and 1.0. In Eq. (4) l is determined by the product of the function $f(\alpha)$ (from Eq. 2) and the function $g(lowest\_value, 1.0)$. Even for a slight disturbance (a value for *lowest_value* just below 1.0) the form is deregulated, and plate-like forms as shown in Figure 9 (form E) are generated. In those plate-like forms new irregularities

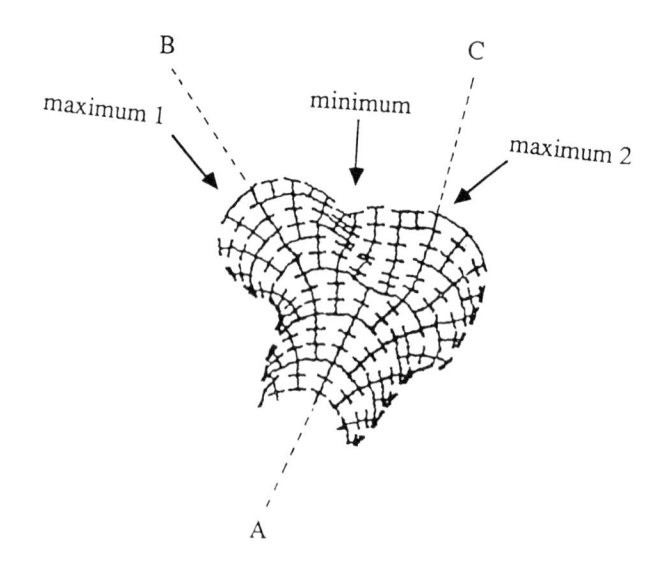

Figure 10 Longitudinal section of a growing object, where two local maxima are arising. The old growth axis (axis A) is replaced by two new growth axes (axes B and C).

Modelling Growth Forms of Sponges With Fractal Techniques 83

are generated during each iteration, which in turn produce new growth axes. In reality, the growth process will be disturbed more on exposed sites; in a turbulent environment, a large variation in growth velocities will occur. Because of this variation, local maxima and minima in growth velocity can be identified on the growing object. The most protruding parts of the sponge will have the highest access to material suspended in the water and will develop the highest growth velocities.

Additional rules for the formation of branches and plates

Two new postprocessing rules are necessary when new growth axes are to be generated during the formation of plates and branches (see below). During this process some active fertile sides will be enclosed by surrounding active fertile sides. In order to prevent collisions, sides of this type are removed from the object. Longitudinal lines, until now ending in a tangential side, will now end somewhere in the object after the enclosed side has been removed. This postprocessing function (*post_nonfitting*) is the reverse of the *post_adding* function. Without the removal of nonfitting elements from the skeleton, a dense accumulation of elements would appear. In reality this accumulation is not found, and elements do not develop in the skeleton when there is not enough space.

Another colliding situation arises when plates or branches intersect each other. A rule preventing intersections (*post_intersection*) changes the status of intersecting active fertile elements into nonactive. A variation on the previous postprocessing function (*post_intersection*) is a rule in which branches are not allowed to intersect at all but are forced by a new rule (*post_avoid*) to keep a certain distance. This rule changes the status of active fertile elements that are approaching collision into nonactive. In reality, water movement and food supply decrease when branches are colliding, and growth will be slowed down or stopped.

Formation of branches

In the growth rules discussed above, the longitudinal length l of newly added elements only depends on the angle α between a growth axis and the active fertile element. In Eq. (5)

$$h(rad\_curv) = 1.0 - (rad\_curv - min\_curv)/(max\_curv - min\_curv)$$
$$min\_curv \leq rad\_curv \leq max\_curv$$
$$h(rad\_curv) = 1.0 \; rad\_curv < min\_curv$$
$$h(rad\_curv) = 0.0 \; rad\_curv > max\_curv$$

$$l = \begin{cases} s \cdot h(rad\_curv) & h(rad\_curv) > inhibition\_level \\ 0.0 & h(rad\_curv) \leq inhibition\_level \end{cases} \quad (5)$$

an alternative growth rule is shown in which the radius of curvature instead of α determines the longitudinal length l. The radius of curvature can be defined as the radius of the circle through three points on the circumference of the growing object. The minimum distance between these points is the length of a tangential side s. The length of the longitudinal side becomes zero as soon as the radius of curvature is larger than a certain (fixed) maximum (max_curv, see Eq. 5). This happens, for example, when three points are situated on one line at the lateral side of a column. l attains the maximum value when the radius of curvature is less then a fixed minimum (min_curv). The radius of curvature is used as an argument ($local\_inf$) for the generator processing function, in which (in this case) Eq. (5) is applied as a growth function. The result of this construction is shown in Figure 9 (form F). The object starts growing as a column, and the longitudinal sides added at the top (close to the growth axis) are equal in size. After some growth stages, the top flattens and the curvature exceeds max_curv. The length l of the longitudinal sides becomes zero, and the original column splits up into two branches. After some stages the regularity of the dichotomous branching pattern is disturbed by intersections.

In reality, plate-like growth forms (Figure 9, form E) are found but not under all circumstances (see Figure 6). The transport of water is sustained only in a limited way by an aquiferous system. Probably transport of water through *Haliclona oculata* is strongly supported by external water movements as well. Under conditions with much water movement plate-like growth forms are possible, whereas under sheltered conditions a decrease of food supply will appear in the tissue unless it is in short distance contact with the environment; a decrease in growth velocity results. This process is modelled with the generator processing function in Eq. (5). In this function the amount of contact with the environment is taken to be proportional to the radius of curvature. As soon as the top is widening too much, the amount of contact with the environment becomes suboptimal and the growth velocity at the top decreases, resulting in a branching object.

A combination of the previous models

A combination of Eqs. (3) and (5) yields a growth rule, in which both the radius of curvature and the angle of a fertile active element with the growth axis are included. The combination of both growth rules can be written as a product of the growth functions, as shown in Eq. (6)

$$l = \begin{cases} s \cdot f(\alpha) \cdot h(rad\_curv) & f(\alpha) \cdot h(rad\_curv) > inhibition\_level \\ 0.0 & f(\alpha) \cdot h(rad\_curv) \leq inhibition\_level \end{cases} \quad (6)$$

Function $f(\alpha)$ is derived from Eq. (3) and function $h(rad\_curv)$ from Eq. (5). In this case $local\_inf$ in the generator processing consists of two components: the rad_curv and α.

In the growing object, small areas with equal-sized longitudinal sides arise at the top. The radius of curvature increases, and the value returned by h decreases. The result is that l decreases and the object starts branching (see Figure 9, form G).

A more evolved object generated with this construction is shown in Plate 18. This object is constructed, with the value $n = 18$, in $f(\alpha)$ in Eq. (6) and obtained after 220 iterations. In this figure only the tangential growth lines are shown. The next object (Plate 19) is generated by disturbing the growth process by multiplying the product $f(\alpha) \cdot h(rad\_curv)$ in Eq. (6) (n = 18) with the function $g()$ (see Eq. 4). The result is that the objects form plate-like branches (like Figure 9, form E). As soon as the radius of curvature of the circumference of the plates exceeds max_curv in Eq. (5) the objects start branching. Here large plates are formed by allowing a larger value for max_curv than applied in Plate 18.

In Figure 9, form G shows the model of a sponge with the highest growth velocities at the protrusions (Figure 9, form C) united with the restrictions of the aquiferous system architecture (Figure 9, form F). The parameter max_curv in Eq. (5) represents the maximum allowed radius of curvature of the surface; its biological interpretation is the minimum amount of contact allowed with the environment before growth velocity decreases and the sponge starts branching. This parameter is closely related to the degree of exposure to water movement, expressed in the model as the degree of disturbance. When $g()$ is increasing the parameter max_curv can also be increased. In the sequence shown in Plates 18 and 19, the parameters max_curv and $g()$ are increased. In Plate 19 an increase in plate forming at the extremities can be seen.

Extension of model H

In Plates 18 and 19 it can be observed that new branches are formed in all directions. In reality the flow and supply of suspended material is higher in the water layers more remote from the bottom. In Eq. (7)

$$k(\beta) = \cos(\beta/2) \qquad 0.0 \leq \beta \leq \pi$$

$$l = \begin{cases} s \cdot f(\alpha) \cdot g() \cdot h(rad\_curv) \cdot k(\beta) & \\ \qquad f(\alpha) \cdot g() \cdot h(rad\_curv) \cdot k(\beta) > inhibition\_level \\ 0.0 & f(\alpha) \cdot g() \cdot h(rad\_curv) \cdot k(\beta) \leq inhibition\_level \end{cases} \qquad (7)$$

this effect is simulated in the growth function. In this function, growth also depends upon the angle β between a growth axis and the y-axis (function $k()$ in Eq. 7). Growth stops as soon as branches start growing downwards and attains a maximum when the growth axis is parallel with the y-axis. The result of this construction (form I in Figure 9) is shown in Plate 20; for this object the same parameters as in Plate 18 were used. Another example is shown in Plate 21, where the same parameters were used as for the object in Plate 19.

Conclusions

With Figure 9 (form I) one can perform a simple 2D simulation of a branching sponge; see also Plates 20 and 21. More irregular forms (as found among sponges several years of age) demand a further extension of the growth model. For example, with a postprocessing rule that represents removing material especially from the protruding parts in order to simulate the process of abrasion.

A major improvement will be the extension of this model to three dimensions. As mentioned in the first section on sponge models, the skeleton of *Haliclona oculata* shows a radial symmetry. Some of the processes (for example, the formation of branches, or plate forming at the extremities of branches) can, because of this symmetry, be described with a 2D model. Other processes, like colliding of branches, can be described adequately only with a 3D model. In Figure 6b it can be seen that the plate-like extremities start growing over each other. In the 2D model (see Plate 21) the colliding plates do not have many possibilities for avoiding each other.

Acknowledgements. The author would like to thank Dr. E.H. Dooijes (Department of Computer Science), Prof. Dr. H.A. Lauwerier (Department of Mathematics), Dr. R.W.M. van Soest (Institute of Taxonomic Zoology), Prof. Dr. J.H. Stock (Institute of Taxonomic Zoology) from the University of Amsterdam and Prof. Dr. J.A.J. Metz (Institute of Theoretical Biology, University of Leiden) for reviewing the manuscript. The author would also like to thank Dr. M.J. de Kluijver (Institute of Taxonomic Zoology), who put a large collection of *Haliclona oculata* at the author's disposal, and Mr. L.A. van der Laan (Institute of Taxonomic Zoology) for preparing the photographs of sponges. The research was done on computer equipment made available by IBM under ACIS contract K513.

REFERENCES

[Aono84]
Aono, M., and Kunii, L., Botanical tree image generation, *IEEE Comput. Graph. and Appl.*, Vol. 8, pp. 10–34, 1984.

[Barn88]
Barnsley, M.F., *Fractals Everywhere*, San Diego: Academic Press, 1988.

[Bell86]
Bell, A.D., The Simulation of Branching Patterns in Modular Organisms, in *The Growth and Form of Modular Organisms*, Harper, J.L., Rosen, B.R., and White, J., Eds., pp. 143–159, London: The Royal Society London, 1986.

[Bloo85]
Bloomenthal, J., Modeling the mighty maple, *Comput. Graph.*, Vol. 19, pp. 305–311, 1985 (SIGGRAPH 85).

[Brad83]
Bradbury, R.H., and Reichelt, R.E., Fractal dimension of a coral reef at ecological scales, *Mar. Ecol. Prog. Ser.*, Vol. 10, pp. 169–172, 1983.

[deRe88]
deReffye, P., Edelin, C., Françon, J., Jaeger, M., and Puech, C., Plant models faithful to botanical structure and development, *Comput. Graph.*, Vol. 22, pp. 151–158, 1988 (SIGGRAPH 88).

[deWe86]
de Weerdt, W.H., A systematic revision of the North-Eastern Atlantic shallow-water Haplosclerida (Porifera, Demospongiae), part III: Chalinidea, *Beaufortia*, Vol. 36, pp. 81–165, 1986.

[Fede88]
Feder, J., *Fractals*, New York: Plenum Press, 1988.

[Grau77]
Graus, R.R., Investigation of coral growth adaptations using computer modeling, in *Proc. Third Internat. Coral Reef Symposium Vol. II*, Taylor, D.L., Ed., pp. 463–469, Miami, Fl: Rosenthiel School of Marine and Atmospheric Sciences, 1977.

[Grau82]
Graus, R.R., and Macintyre, I.G., Variation in growth forms of the reef coral *Montastrea annularis* (Ellis and Solander): a quantitative evaluation of growth response to light distribution using computer simulation, *Smithson. Contr. Mar. Sci.*, Vol. 12, pp. 441–464, 1982.

[Hart58]
Hartman, W.D., Natural history of the marine sponges of southern New England, *Bull. Peabody Mus. Nat. Hist.*, Vol. 12, pp. 1–155, 1958.

[Hoge74]
Hogeweg, P., and Hesper, B., A model study on biomorphological description, *Pattern Recoq.*, Vol. 6, pp. 165–179, 1974.

[Hopg83]
Hopgood, F.R.A., Duce, D.A., Gallop, J.R., and Sutcliffe, D.C., *Introduction to the Graphical Kernel system (GKS)*, London: Academic Press, 1983.

[Jack79]
Jackson, J.B.C., Morphological Strategies of Sessile Animals, in *Biology and Systematics of Colonial Organisms Volume II*, Larwood, C., and Rosen, B.R., Eds., London: Academic Press, pp. 499–555, 1979 (Proceedings of international symposium held at University of Durham, Systematics Association.

[Kaan87]
Kaandorp, J.A., Interactive generation of fractal objects, in *Proc. European Comput. Graph. Conf.*, Marechal, G., Ed., pp. 181–196, Amsterdam: North-Holland, 1987.

[Kawa82]
Kawaguchi, Y., A morphological study of the form of nature, *Comput. Graph.*, Vol. 16, pp. 223–232, 1982 (SIGGRAPH 82).

[Kell87]
Keller, J.M., Characteristics of natural scenes related to fractal dimension, *IEEE Trans. Patt. Anal. Mach. Intell.*, Vol. 9, pp. 621–627, 1987.

[Lewi85]
Lewis, M., and Rees, D.C., Fractal surfaces of proteins, *Science*, Vol. 230, pp. 1163–1165, 1985.

[Lind68]
Lindenmayer, A., Mathematical models for cellular interactions in development, *J. Theor. Biol.*, Vol. 18, pp. 280–299, 1968.

[Mand83]
Mandelbrot, B.B., *The Fractal Geometry of Nature*, New York: W.H. Freeman, 1983.

[Meak86]
Meakin, P., A new model for biological pattern formation, *J. Theor. Biol.*, Vol. 118, pp. 101–113, 1986.

[Mors85]
Morse, D.R., Lawton, J.H., Dodson, M.M., and Williamson, M.H., Fractal dimension of vegetation and the distribution of arthropod body length, *Nature*, Vol. 314, pp. 731–733, 1985.

[Oppe86]
Oppenheimer, P.E., Real time design and animation of fractal plants and trees, *Comput. Graph.*, Vol. 20, pp. 55–64, 1986 (SIGGRAPH 86).

[Pent84]
Pentland, A.P., Fractal-based description of natural scenes, *IEEE Trans. Patt. Anal. Mach. Intell.*, Vol. 6, pp. 661–674, 1984.

[Prus88]
Prusinkiewicz, P., Lindenmayer, A., and Haman, J., Developmental models of herbaceous plants for computer imagery purposes, *Comput. Graph.*, Vol. 22, pp. 141–150, 1988 (SIGGRAPH 88).

[Shue86]
Shuey, D., Bailey, D., and Morrissey, T.P., Phigs: a standard, dynamic, interactive graphics interface, *IEEE Comput. Graph. and Appl.*, Vol. 6, pp. 50–57, 1986.

[Stan87]
Stanley, H.E., and Ostrowsky, N., *On Growth and Form: Fractal and Non-fractal Patterns in Physics*, Boston: Martinus Nijhoff, 1987.

[Wied77]
Wiedenmayer, F., *Shallow-water Sponges of the Western Bahamas*, Basel: Birkhauser Verlag, 1977.

[Witt81]
Witten, T.A., and Sander, L.M., Diffusion-limited aggregation, a kinetic critical phenomenon, *Phys. Rev. Lett.*, Vol. 47, pp. 1400–1403, 1981.

Random Fractals in Image Synthesis

Dietmar Saupe

Abstract

Random fractals are the method of choice when it comes to modelling landscapes, clouds and other natural phenomena for the purpose of computer image synthesis. This paper describes the fundamentals of random fractals and some of the basic methods for their generation. This includes the midpoint displacement and the spectral synthesis methods as well as a functional-based approach. Methods using stochastic noise synthesis are not covered.

Introduction

The fascination that surrounds fractals seems to have two roots. First, fractals are very suitable for simulating many natural phenomena. Stunning pictures have already been produced, and it will not take very long until an uninitiated observer will no longer be able to tell whether a given scene is natural or just computer simulated. The other reason for the fascination is that fractals are simple to generate on computers. In order to generate a fractal one does not have to be an expert in some involved theory. More importantly, the complexity of a fractal, when measured in terms of the length of the shortest computer program that can generate it, is very small.

Given that random fractals have infinite detail at all scales, it follows that a complete computation of a fractal is impossible. So approximation of fractals down to some finite precision has to suffice. The desired level of resolution is naturally given by constraints, such as the number of pixels on the available graphics display, or the amount of computer time that one is willing to spend. The algorithms presented here fall into several categories.

(C1) An approximation of a random fractal with some resolution is used as input, and the algorithm produces an improved approximation with resolution increased by a certain factor. This process is repeated with outputs used as new inputs, until the desired resolution is achieved. In some cases the procedure can be formulated as a recursion. The *midpoint displacement methods* are examples.

(C2) Only one approximation of a random fractal is computed, namely for the final resolution. Pictures for different resolutions are possible but require that most of the computations be redone. The *Fourier filtering method* is an example.

(C3) In the third approach the approximation of a fractal is obtained via an iteration. After each step the approximation is somewhat improved, but the spatial resolution does not necessarily increase by a constant factor in each iteration. Here the allowed computer time determines the quality of the result. The *random cut method* described below, or the computation of Julia sets as preimages of repellers, are examples.

(C4) It is possible to approximate fractals simply by evaluating a function at points of interest. This is called *functional-based modelling*. The *Mandelbrot-Weierstrass function* and *rescale-and-add method* are examples.

(C5) *Stochastic noise synthesis* is similar to the displacement methods. In this method arbitrary points (not only midpoints) may be added to an approximation of a random fractal, taking into account correlations not only with the immediate neighboring points by also with points farther away. Moreover, similar to the Fourier filtering method, other types of stochastic processes besides those used for random fractals may be simulated. In this paper these noise synthesis methods are not covered (see [Fell85; Lewi87; Lewi89; Rüme90]).

Almost all of the algorithms contained in this paper are based on methods and ideas discussed in Peitgen [Peit88], as well as in Voss [Voss85] and ultimately in Mandelbrot [Mand82a], where some background information and further references can be found.

It must be acknowledged that the generation of data for a fractal alone is not sufficient to obtain a good picture. It is likely that computer programmers will spend far more time writing software that can render, e.g., a fractal landscape with some reasonable coloring and shading (if that work has not already been done) than for the coding of fractal data generation routines. In any case the computer will surely use much more CPU time for rendering than for data generation. This is a good reason not to be satisfied with poor approximations of fractals due to insufficient algorithms. It is beyond the scope of this paper to explain the rendering aspects of random fractals. There are some references included in the list at the end that deal with such questions, e.g., [Four85; Harr87; Hear86; Mill86; Musg89a]

In the following sections we describe a number of algorithms with increasing complexity. Some of them are documented in the form of pseudocode which we hope is self-explanatory. This code is included to clarify the methods and therefore is not optimized for speed.

These notes are basically a condensed version of Chapter 2 of *The Science of Fractal Images* [Peit88] and the paper 'Point evaluation of multi-variable random fractals' [Saup89a]. Some heuristics, proofs and further implementation details missing here may be found in these references.

First Case Study: One-dimensional Brownian Motion

DEFINITIONS

Brownian motion in one variable constitutes the simplest random fractal, and it is at the heart of many of the following generalizations. Small particles of solid matter suspended in a liquid can be seen under a microscope to move about in an irregular and erratic way. This was observed by the botanist R. Brown around 1827. When restricting the motion to one dimension we obtain a random process $X(t)$, i.e., a function X of a real variable t (time) whose values are random variables $X(t_1)$, $X(t_2)$, etc.

The increment
$$X(t_2) - X(t_1) \tag{1}$$
has Gaussian distribution, and the mean square increments have a variance proportional to the time differences. Thus
$$E\left[|X(t_2) - X(t_1)|^2\right] \propto |t_2 - t_1| \tag{2}$$

Here E denotes the mathematical expectation of a random variable or, in other words, the average over many samples. We say that the increments of X are *statistically self-similar* in the sense that

$$X(t_0 + t) - X(t_0) \quad \text{and} \quad \frac{1}{\sqrt{r}}\left(X(t_0 + rt) - X(t_0)\right)$$

have the same finite dimensional joint distribution functions for any t_0 and $r > 0$. If we take for convenience $t_0 = 0$ and $X(t_0) = 0$, then this means that the two random functions

$$X(t) \quad \text{and} \quad \frac{1}{\sqrt{r}} X(rt)$$

are statistically indistinguishable. The second one is just a *properly rescaled* version of the first (see Figure 1). Thus, if we accelerate the process $X(t)$ by a factor of 16, for example, then we can divide $X(16t)$ by 4 to obtain the same Brownian motion that we started with.

INTEGRATING WHITE NOISE

The integral of uncorrelated white Gaussian noise W satisfies Eq. (1) and Eq. (2)
$$X(t) = \int_{-\infty}^{t} W(s)\,ds \tag{3}$$

The random variables $W(t)$ are uncorrelated and have the same normal distribution $N(0, 1)$. Moreover, the graph of a sample of Brownian motion $X(t)$ has a fractal dimension of 1.5, and the intersection of the graph with a horizontal line has a dimension of 0.5 (see Figure 2).

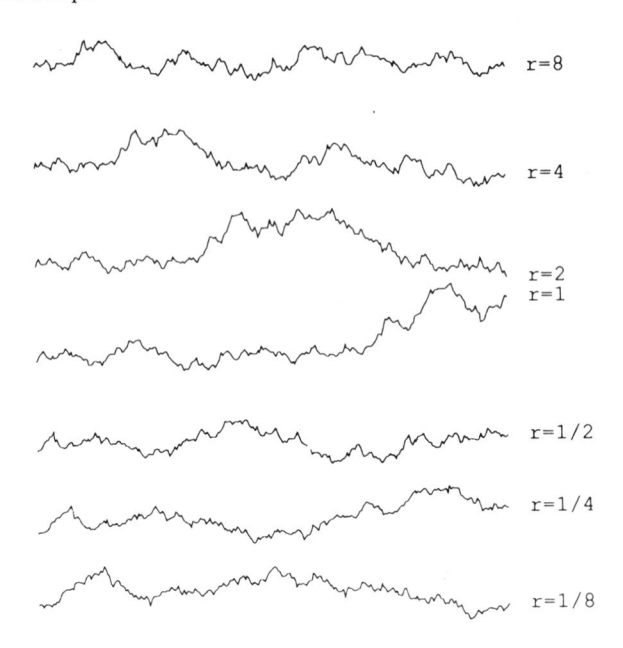

Figure 1. Properly rescaled Brownian motion. The graph in the center shows a small section of Brownian motion $X(t)$. In the other graphs the properly rescaled random functions of the form $r^{-1/2}X(rt)$ are displayed. The scaling factor r ranges from $r = 1/8$ to $r = 8$, corresponding to expanding and contracting the original function in the time direction. Note that the visual appearance of all samples is the same.

Random Midpoint Displacement Method

Another straightforward method to produce Brownian motion is random midpoint displacement (see Figure 3). If the process is to be computed for times t between 0 and 1, one starts by setting $X(0) = 0$ and selecting $X(1)$ as a sample of a Gaussian random variable with mean 0 and variance σ^2. Then $\text{var}\,(X(1) - X(0)) = \sigma^2$ and we expect

$$\text{var}\,(X(t_2) - X(t_1)) = |t_2 - t_1|\sigma^2 \tag{4}$$

for $0 \leq t_1 \leq t_2 \leq 1$. We set $X(1/2)$ to be the average of $X(0)$ and $X(1)$ plus some Gaussian random offset D_1, with mean 0 and variance Δ_1^2. Then

$$X(\tfrac{1}{2}) - X(0) = \frac{1}{2}(X(1) - X(0)) + D_1$$

and thus $X(1/2) - X(0)$ has mean value 0. The same holds for $X(1) - X(1/2)$. Secondly, for Eq. (4) to be true we must require

$$\text{var}\,(X(\tfrac{1}{2}) - X(0)) = \frac{1}{4}\text{var}\,(X(1) - X(0)) + \Delta_1^2 = \frac{1}{2}\sigma^2$$

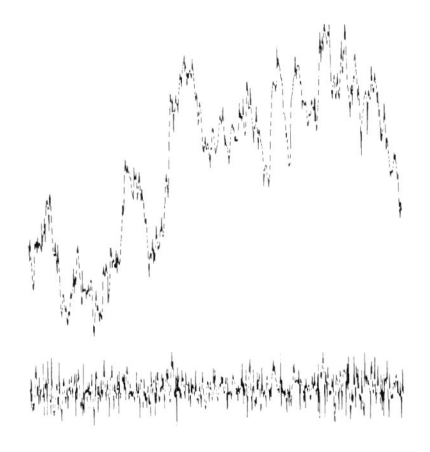

Figure 2. Brownian motion in one dimension. The lower graph shows a sample of Gaussian white noise; its time integral yields the Brownian motion above it.

Therefore
$$\Delta_1^2 = \frac{1}{4}\sigma^2$$
In the next step we proceed in the same fashion setting
$$X(\tfrac{1}{4}) - X(0) = \frac{1}{2}\bigl(X(0) + X(\tfrac{1}{2})\bigr) + D_2$$
and observe that again the increments in X, here $X(1/2) - X(1/4)$ and $X(1/4) - X(0)$ are Gaussian and have mean 0. So we must choose the variance Δ_2^2 of D_2 such that
$$\text{var}\bigl(X(\tfrac{1}{4}) - X(0)\bigr) = \frac{1}{4}\text{var}\bigl(X(\tfrac{1}{2}) - X(0)\bigr) + \Delta_2^2 = \frac{1}{4}\sigma^2$$
holds, i.e.,
$$\Delta_2^2 = \frac{1}{8}\sigma^2$$
We apply the same idea to $X(3/4)$ and continue to finer resolutions yielding
$$\Delta_n^2 = \frac{1}{2^{n+1}}\sigma^2$$

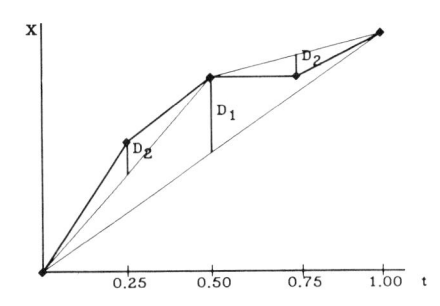

Figure 3. Midpoint displacement: the first two stages in the midpoint displacement technique as explained in the text.

94   Dietmar Saupe

as the variance of the displacement D_n. Thus, corresponding to time differences $\Delta t = 2^{-n}$ we add a random element of variance, $2^{-(n+1)}\sigma^2$, which is proportional to Δt, as expected. Figure 4 shows the result of the first eight stages in the algorithm.

INDEPENDENT JUMPS

Our last algorithm for Brownian motion falls into category C3. We may interpret Brownian motion as the cumulative displacement of a series of independent jumps, i.e., an infinite sum of functions

$$J_i(t) = A_i \beta(t - t_i)$$

where
$$\beta(t) = \begin{cases} 0 & \text{if } t < 0 \\ 1 & \text{if } t \geq 0 \end{cases}$$

and A_i, t_i are random variables with Gaussian and Poisson distributions, respectively. This approach generalizes to circles and spheres. For circles we take time as 2π periodic and arrive at

$$X(t) = \sum_{i=0}^{\infty} A_i \overline{\beta}(t - t_i)$$

where
$$\overline{\beta}(t) = \begin{cases} 0 & \text{if } t \geq \pi \pmod{2\pi} \\ 1 & \text{if } t < \pi \pmod{2\pi} \end{cases} \tag{5}$$

The A_i are identically distributed Gaussian random variables, and the t_i are uniformly distributed random variables with values between 0 and 2π. Each term in the sum in Eq. (5) adds random displacement on half of the circle.

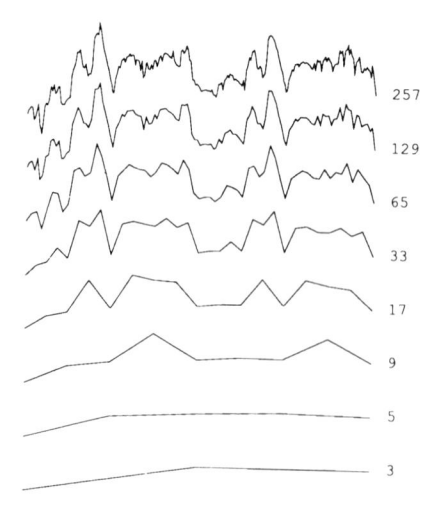

Figure 4. Brownian motion via midpoint displacement method. Eight intermediate stages of the algorithm *MidPointBM()* are shown depicting approximations of Brownian motion using up to 3, 5, 9, ..., 257 points, i.e., the parameter maxlevel was set to 2, 3, 4, ..., 8.

ALGORITHM	MidPointBM(X, maxlevel, sigma, seed)	
Title	Brownian motion via midpoint displacement	
Arguments	X[]	real array of size $2^{maxlevel} + 1$
	maxlevel	maximal number of recursions
	sigma	initial standard deviation
	seed	seed value for random number generator
Globals	delta[]	array holding standard deviations Δ_i
Variables	i, N	integers

BEGIN
   InitGauss(seed)
   FOR i := 1 TO maxlevel DO
      delta[i] := sigma*power(0.5, (i+1)/2)
   END FOR
   N = power(2, maxlevel)
   X[0] := 0
   X[N] := sigma*Gauss()
   MidPointRecursion(X, 0, N, 1, maxlevel)
END

Of course, we can also use midpoint displacement to obtain Brownian motion on a circle. We would just have to require that $X(0) = X(1)$ in order to maintain continuity. However, the midpoint method does not generalize to spheres, whereas the random cut method does: In each step a great circle of the sphere

ALGORITHM	MidPointRecursion(X, index0, index2, level, maxlevel)	
Title	Recursive routine called by MidPointBM()	
Arguments	X[]	real array of size $2^{maxlevel} + 1$
	index0	lower index in array
	index2	upper index in array
	level	depth of the recursion
	maxlevel	maximal number of recursions
Globals	delta	array holding standard deviations Δ_i
Variables	index1	midpoint index in array

BEGIN
   index1 := (index0 + index2)/2
   X[index1] := 0.5*(X[index0] + X[index2]) + delta[level]*Gauss()
   IF (level < maxlevel) THEN
      MidPointRecurs(X, index0, index1, level+1, maxlevel)
      MidPointRecurs(X, index1, index2, level+1, maxlevel)
   END IF
END

is picked at random, and one of the two half-spheres is displaced by an amount determined by a Gaussian random variable. The famous pictures of the planets in Mandelbrot [Mand82a], Voss [Voss85] and Peitgen [Peit88] were produced in this way by Voss.

Fractional Brownian Motion: Approximation by Spatial Methods

DEFINITIONS

In the last section we studied random processes $X(t)$ with Gaussian increments and
$$\text{var}(X(t_2) - X(t_1)) \propto |t_2 - t_1|^{2H} \tag{6}$$
where $H = 1/2$. The generalization to parameters $0 < H < 1$ is called *fractional Brownian motion* (fBm, see [Mand68], [Mand82a]). As in the case of ordinary Brownian motion, we say that the increments of X are *statistically self-similar with parameter H*, in other words

$$X(t_0 + t) - X(t_0) \quad \text{and} \quad \frac{1}{r^H}\left(X(t_0 + rt) - X(t_0)\right)$$

have the same finite-dimensional joint distribution functions for any t_0 and $r > 0$. If we again use for convenience $t_0 = 0$ and $X(t_0) = 0$, the two random functions

$$X(t) \quad \text{and} \quad \frac{1}{r^H} X(rt) \tag{7}$$

are statistically indistinguishable. Thus, 'accelerated' fractional Brownian motion $X(rt)$ is *properly rescaled* by dividing amplitudes by r^H.

Let us visualize this important fact (see Figures 1 and 5). If we set $H = 1/2$ we obtain the usual Brownian motion of the last section. For $H = 0$ we get a completely different behavior of X: We can expand or contract the graph of X in the t direction by any factor, and the process will still 'look' the same. This clearly says that the graph of a sample of X must densely fill up a region in the plane. In other words, its fractal dimension is 2. The opposite case is given by the parameter $H = 1$. There we must compensate for an expansion of the graph in the t direction by also multiplying the amplitudes by the same factor. It is easy to give an argument showing that the fractal dimension of this graph must be $2 - H = 1$. In fact, graphs of samples of fBm have a fractal dimension of $2 - H$ for $0 < H < 1$. The parameter H thus describes the 'roughness' of the function at small scales.

Fractional Brownian motion can be divided into three quite distinct categories: $H < 1/2$, $H = 1/2$ and $H > 1/2$. The case $H = 1/2$ is the ordinary Brownian motion which has independent increments, i.e., $X(t_2) - X(t_1)$ and $X(t_3) - X(t_2)$, with $t_1 < t_2 < t_3$, are independent in the sense of probability

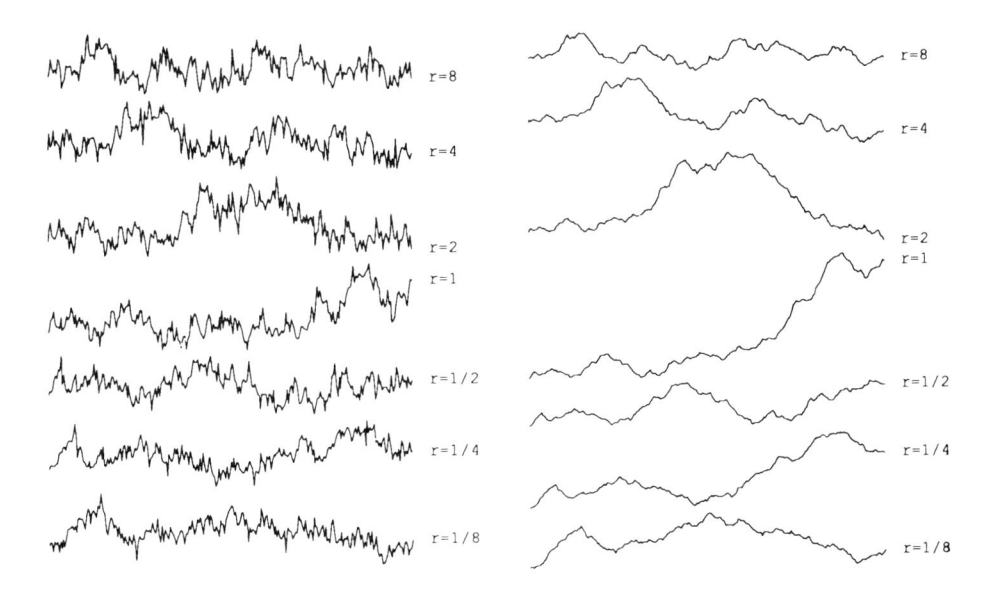

Figure 5. Properly rescaled fractional Brownian motion. The two sets of curves show properly rescaled fractional Brownian motion for parameters $H = 0.2$ (left) and $H = 0.8$ (right). The graphs in the center show small sections of fractional Brownian motion $X(t)$. In the other graphs, the properly rescaled random functions $r^{-H}X(rt)$ are displayed. The scaling factor r ranges from $r = 1/8$ to $r = 8$, corresponding to expansion and contraction of the original function in the time direction. Compare with Figure 1 for $H = 0.5$.

theory; their correlation is 0. For $H > 1/2$ there is a positive correlation between these increments, i.e., if the graph of X is increasing for some t_0, then it tends to continue to increase for $t > t_0$. For $H < 1/2$ the opposite is true. There is a negative correlation of increments, and the curves seem to oscillate more erratically.

Midpoint Displacement Methods

For the approximation of fBm the approach taken by the midpoint method can formally be extended to suit parameters $H \neq 1/2$ (see Figure 6 and the algorithm *MidPointFM1D()*). Here we aim for the equivalent of Eq. (4)

$$\text{var}(X(t_2) - X(t_1)) = |t_2 - t_1|^{2H} \sigma^2$$

Using the same line of thought as in the last section, we arrive at midpoint displacements D_n that have variances

$$\Delta_n^2 = \frac{\sigma^2}{(2^n)^{2H}}(1 - 2^{2H-2}) \qquad (8)$$

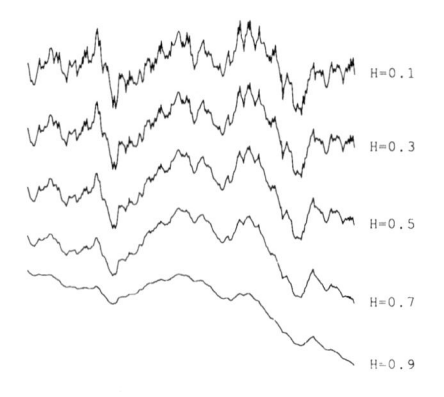

Figure 6. Fractal motion generated by displacement techniques. The curves are results of the algorithm *MidPointFM1D()* for various parameters H.

Therefore the only changes required in the algorithm *MidPointBM()* occur in the computation of Δ_n accommodating $H \neq 1/2$.

It has been shown that the above midpoint displacement technique does not yield true fBm for $H \neq 1/2$ [Mand82b]. In fact, although

$$\mathrm{var}\left(X(\tfrac{1}{2}) - X(0)\right) = \mathrm{var}\left(X(1) - X(\tfrac{1}{2})\right) = \left(\frac{1}{2}\right)^{2H} \sigma^2$$

we do not have

$$\mathrm{var}\left(X(\tfrac{3}{4}) - X(\tfrac{1}{4})\right) = \left(\frac{1}{2}\right)^{2H} \sigma^2$$

ALGORITHM	MidPointFM1D(X, maxlevel, sigma, H, seed)	
Title	One-dimensional fractal motion via midpoint displacement	
Arguments	X[]	real array of size $2^{maxlevel} + 1$
	maxlevel	maximal number of recursions
	sigma	initial standard deviation
	H	$0 < H < 1$ determines fractal dimension $D = 2 - H$
	seed	seed value for random number generator
Globals	delta[]	array holding standard deviations Δ_i
Variables	i, N	integers

```
BEGIN
    InitGauss(seed)
    FOR i := 1 TO maxlevel DO
        delta[i] := sigma*power(0.5, i*H)*sqrt(1−power(2, 2*H-2))
    END FOR
    N = power(2, maxlevel)
    X[0] := 0
    X[N] := sigma*Gauss()
    MidPointRecursion(X, 0, N, 1, maxlevel)
END
```

as we would like. Thus, this process does not have stationary increments; the times t are not all statistically equivalent. This defect causes the graphs of X to show some visible traces of the first few stages in the recursion. In the two-dimensional extension of this algorithm one obtains results which may exhibit some disturbing creases along straight lines, which are related to the underlying grid of points. Nevertheless, this is still a useful algorithm for many purposes. It became most popular after its appearance in Fournier [Four82] and subsequently in some science magazines. Recently it has also been included as a part of some computer graphics text books ([Harr87],[Hear86]).

One approach to deal with the lack of stationarity of the random functions generated by the midpoint displacement technique is to interpolate the midpoints in the same way but to then add a displacement D_n of a suitable variance to all of the points and not just to the midpoints. This seems natural, as one would reiterate a measurement at all points in a graph of fBm when a device is used that allows measurements at a smaller sampling rate Δt and a smaller spatial resolution ΔX. This method is called *successive random addition*. The extra amount of work involved as compared with the midpoint method is tolerable; about twice as many displacements are necessary. The actual formula for the variances of the displacements as used in the various stages of this algorithm (see algorithm *AdditionsFM1D()*) are derived as a special case of the method discussed in the next section.

Displacing Interpolated Points

In the midpoint displacement method and the successive random addition method, the resolution Δt is improved by a factor of $r = 1/2$ in each stage. We can modify the second of the above methods to accommodate other factors $0 < r < 1$. For this purpose one would interpolate $X(t)$ at times $t_i = ir\Delta t$ from the samples one already has from the previous stage at a sampling rate of Δt. Then a random element D_n would be added to all of the interpolated points. In agreement with the requirements for the mean square increments of X, we have to prescribe a variance proportional to $(r^n)^{2H}$ for the Gaussian random variable D_n

$$\Delta_n^2 \propto (r^n)^{2H} \tag{9}$$

The additional parameter r will change the appearance of the fractal; the feature of the fractal controlled by r has been termed the *lacunarity* (see Figure 7). In the algorithm below we use a simple linear interpolation. The reason we include yet another method here is that it generalizes to two, three or more dimensions in a very straightforward way, while the other methods are harder to convert.

Following the ideas for midpoint displacement for Brownian motion, we set $X(0) = 0$ and select $X(1)$ as a sample of a Gaussian random variable with variance σ^2. Then we can deduce in the same fashion as before that

$$\Delta_n^2 = \frac{1}{2}(1 - r^{2-2H})(r^n)^{2H}\sigma^2$$

```
ALGORITHM  AdditionsFM1D(X, maxlevel, sigma, H, seed)
Title      One-dimensional fractal motion via successive random additions

Arguments   X[]       real array of size 2^maxlevel + 1
            maxlevel  maximal number of recursions
            sigma     initial standard deviation
            H         0 < H < 1 determines fractal dimension D = 2 − H
            seed      seed value for random number generator
Globals     delta[]   array holding standard deviations Δ_i
Variables   i, N, d, D integers
            level     integer

BEGIN
   InitGauss(seed)
   FOR i := 1 TO maxlevel DO
       delta[i] := sigma*power(0.5, i*H)*sqrt(0.5)*sqrt(1-power(2,2*H-2))
   END FOR
   N = power(2, maxlevel)
   X[0] := 0
   X[N] := sigma*Gauss()
   D := N
   d := D/2
   level := 1
   While (level ≤ maxlevel) DO
       For i := d TO N-d STEP D DO
           X[i] := 0.5*(X[i-d] + X[i+d])
       END FOR
       For i := 0 TO N STEP d DO
           X[i] := X[i] + delta[level]*Gauss( )
       END FOR
       D :+ D/2
       d := d/2
       level := level + 1
   END WHILE
END
```

A little difficulty arises, since we cannot work with a continuous variable $X(t)$ in a computer program (X is stored as an array). Thus, generally one cannot maintain the constant factor of r reducing the resolution Δt in each step. Let us assume for simplicity that we have some $X(0)$ and $X(T)$ already computed, where $T \approx r^{n-1}$. Then $X(t)$, with $t \approx r^n$, is first interpolated via

$$X(t) = X(0) + \frac{t}{T}(X(T) - X(0))$$

Then all points $X(0), X(t), X(2t)$, etc. are displaced by samples of a Gaussian random variable D with variance Δ^2. The old values $X(0)$ and $X(T)$ satisfied

$$\text{var}(X(T) - X(0)) = T^{2H}\sigma^2$$

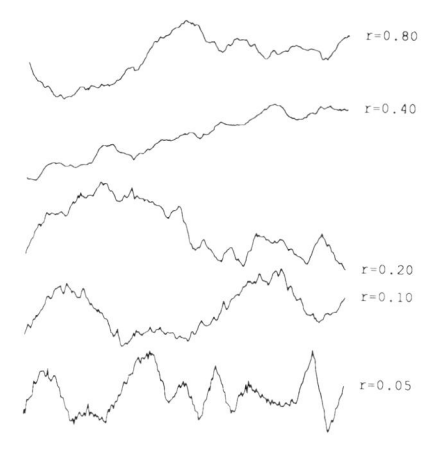

Figure 7. Fractal Brownian motion with varying lacunarity. The algorithm *InterpolatedFM1D()* produced these curves with a parameter $H = 0.8$. The other parameter r is the scaling factor in the method. $r = 0.1$ means that in each stage we have $1/r = 10$ times as many points as in the previous stage. Especially for low r we see a pronounced effect of these high scaling ratios as a modulation remaining from the first one or two stages. This is called the lacunarity.

and we have to select the variance Δ^2 such that

$$\text{var}(X(t) - X(0)) = t^{2H}\sigma^2$$

holds. Therefore we require

$$t^{2H}\sigma^2 = \left(\frac{t}{T}\right)^2 T^{2H}\sigma^2 + 2\Delta^2$$

where the first term of the sum stands for the variance already contained in the interpolation, and the other represents the variance due to the perturbations in both $X(0)$ and $X(t)$. Thus

$$\Delta^2 = \frac{1}{2}\sigma^2\left(1 - \left(\frac{t}{T}\right)^{2-2H}\right)t^{2H}$$

To apply this formula to the case of random successive additions we set $T = 1/2^{n-1}$ and $t = 1/2^n$, and we obtain

$$\Delta^2 = \frac{1}{2}\sigma^2\left(1 - \frac{1}{2^{2-2H}}\right)\frac{1}{2^{2Hn}}$$

which is, as expected, the quantity Δ_n^2 used in the algorithm *AdditionsFM1D()*, where the ratio r is $1/2$.

Fractional Brownian Motion : Approximation by Spectral Synthesis

THE SPECTRAL EXPONENT β IN FRACTIONAL BROWNIAN MOTION

The underlying idea of spectral synthesis is that a prescription of the right kind of spectral density $S(f)$ will give rise to fBm with an exponent $0 < H < 1$.

If the random function $X(t)$ contains equal power for all frequencies f, this process is called white noise, in analogy with the white light made up of radiations of all visible wave lengths. If $S(f)$ is proportional to $1/f^2$ we obtain the usual brown noise, or Brownian motion. In general, a process $X(t)$ with a spectral density proportional to $1/f^\beta$ corresponds to fBm with $H = (\beta-1)/2$

$$S(f) \propto \frac{1}{f^\beta} \sim \text{fBm with } \beta = 2H+1 \tag{10}$$

This relation can be verified based on the scaling laws of fBm [Peit88]. Choosing β between 1 and 3 will generate a graph of fBm with a fractal dimension of

$$D_f = 2 - H = \frac{5-\beta}{2} \tag{11}$$

THE FOURIER FILTERING METHOD

For a practical algorithm we have to translate the above into conditions on the coefficients a_k of the discrete Fourier transform

$$\overline{X}(t) = \sum_{k=0}^{N-1} a_k e^{2\pi i k t} \tag{12}$$

The coefficients a_k are in a 1:1 correspondence with the complex values $\overline{X}(t_k)$, $t_k = k/N$, $k = 0, 1, \ldots, N-1$. The condition to be imposed on the coefficients in order to obtain $S(f) \propto 1/f^\beta$ now becomes

$$Eb\left(|a_k|^2\right) \propto \frac{1}{k^\beta} \tag{13}$$

since k essentially denotes the frequency in Eq. (12). Equation (13) holds for $0 < k < N/2$. For $k \geq N/2$ we must have $a_k = \overline{a_{N-k}}$ because $\overline{X}$ is a real function. The method simply consists of randomly choosing coefficients subject to the expectation in Eq. (13) and then computing the inverse Fourier transform to obtain X in the time domain. Since the process X need only have real values, it is actually sufficient to sample real random variables A_k and B_k under the constraint

$$E(A_k^2 + B_k^2) \propto \frac{1}{k^\beta}$$

and then setting

$$\overline{X}(t) = \sum_{k=1}^{N/2} (A_k \cos kt + B_k \sin kt) \tag{14}$$

In contrast to some of the previous algorithms discussed, this method is not recursive and does not proceed in stages of increasing spatial resolution. We may, however, interpret the addition of more and more random Fourier coefficients a_k, satisfying Eq. (13) as a process of adding higher frequencies, thus increasing the resolution in the frequency domain.

The advantage of this straightforward method is that it is the purest interpretation of the concept of fractional Brownian motion. Artifacts such as those occurring in the midpoint displacement methods are not apparent. However, due to the nature of Fourier transforms the generated samples are periodic. This is sometimes annoying, and in this case one can compute twice or four times as many points as actually needed and then discard a part of the sequence.

The algorithm for the inverse Fourier transformation *InvFFT* is not included with the pseudocode *SpectralSynthesisFM1D()*. It should compute the sums in Eq. (14) from the given coefficients in A and B. Usually fast Fourier transforms are employed. They require $O(N \log N)$ operations per transform.

Extensions to higher dimensions

DEFINITIONS

In this section we discuss how one can generalize the displacement methods and the spectral synthesis methods to two and three dimensions.

| ALGORITHM | SpectralSynthesisFM1D(X, N, H, seed) | |
Title	Fractal motion using Fourier Filtering method	
Arguments	X[ ]	array of reals of size N
	N	size of array X
	H	$0 < H < 1$ determines fractal dimension $D = 2 - H$
	seed	seed value for random number generator
Globals	Arand	rand( ) returns values between 0 and Arand
Variables	i	integer
	beta	exponent in the spectral density function $(1 < \beta < 3)$
	rad, phase	polar coordinates of Fourier coefficient
	A[ ], B[ ]	real and imaginary parts of Fourier coefficients
Subroutines	InvFFT	fast inverse Fourier transform in 1 dimension

```
BEGIN
    InitGauss(seed)
    beta := 2*H + 1
    FOR i := 0 TO N/2-1  DO
        rad := power(i+1, -beta/2)*Gauss()
        phase := 2*3.141592*rand()/Arand
        A[i] := rad*cos(phase)
        B[i] := rad*sin(phase)
    END FOR
    InvFFT(A, B, X, N/2)
END
```

The generalization of fractional Brownian motion itself is straightforward. It is a multidimensional process (a random field) $X(t_1, t_2, \ldots, t_n)$ with the properties

(a) The increments $X(t_1, t_2, \ldots, t_n) - X(s_1, s_2, \ldots, s_n)$ are Gaussian with mean 0.
(b) The variance of the increments $X(t_1, t_2, \ldots, t_n) - X(s_1, s_2, \ldots, s_n)$ depends only on the distance

$$\sqrt{\sum_{i=1}^{n}(t_i - s_i)^2}$$

and, in fact, is proportional to the $2H$ power of the distance, where the parameter H again satisfies $0 < H < 1$. Thus,

$$E\left(|X(t_1, t_2, \ldots, t_n) - X(s_1, s_2, \ldots, s_n)|^2\right) \propto \left(\sum_{i=1}^{n}(t_i - s_i)^2\right)^H \quad (15)$$

The random field X again has stationary increments and is isotropic, i.e., all points $(t_1, t_2, \ldots, t_n)$ and all directions are statistically equivalent. In the frequency domain we have for the spectral density

$$S(f_1, \ldots, f_n) \propto \frac{1}{\left(\sqrt{\sum_{i=1}^{n} f_i^2}\right)^{2H+n}} \quad (16)$$

This ensures that X restricted to any straight line will be a $1/f^\beta$ noise corresponding to $2H = \beta - 1$ [Voss85]. In analogy with Eq. (10) the fractal dimension of the graph of a sample of $X(t_1, t_2, \ldots, t_n)$ is

$$D = n + 1 - H \quad (17)$$

DISPLACEMENT METHODS

The midpoint displacement methods can work with square lattices of points. If the mesh size δ denotes the resolution of such a grid, we obtain another square grid of resolution $\delta/\sqrt{2}$ by adding the midpoints of all squares. Of course, the orientation of the new square lattice is rotated by 45 degrees. Again adding the midpoints of all squares gives us the next lattice with the same orientation as the first one, and the resolution is now $\delta/2$ (see Figure 8). In each stage we thus scale the resolution with a factor of $r = 1/\sqrt{2}$, and in accordance with Eq. (15) we add random displacements using a variance which is r^{2H} times the variance of the previous stage. If we assume that the data on the four corner points of the grid carry mean square increments of σ^2, then at stage n of the process we must add a Gaussian random variable of variance $\sigma^2 r^{2Hn} = \sigma^2 (1/2)^{nH}$. For the usual midpoint method random elements are added only to the new midpoints in each stage, whereas in the random addition method we add displacements at all points. Thus we can unify both methods in just one algorithm.

In Figure 9 we show a topographical map of a random fractal landscape which was generated using the algorithm *MidPointFM2D()*. Random additions were used, $H = 0.8$. At a resolution of 65×65 points (*maxlevel* $= 6$) a total of 4225 data points were generated. The heights of the surface were scaled to the range from -10000 to $+10000$ and then handed to a contour line program which produced the maps. The parts of the surfaces that have a negative height are assumed to be submerged under water and are not displayed. The fractal dimension of these surfaces is $3 - H$. A rendering of a perspective view of these landscapes is also included in Figure 9

We also remark that one does not necessarily have to work with rectangular grids. Some authors also consider triangular subdivisions (see [Four82, Harr87, Hear86, Mill86]). For a detailed discussion of displacement techniques, their history and possible extensions, we refer to [Mand88].

From the above section on displacing interpolated points the algorithm *InterpolatedFM1D()* which implements variable scaling factors of resolutions can be modified to compute approximations of fBm in two or three (or even higher) dimensions. If d denotes this dimension, we have to fill an array of size N^d elements subject to Eq. (15). These points are evenly spaced grid points in a unit cube, thus the final resolution (grid size) is $1/(N-1)$. At a particular stage of this algorithm we have an approximation at resolution $1/(M-1)$ given in the form of an array of size M^d, with $M < N$. The next stage is approached in two steps: (1) If $0 < r < 1$ denotes the factor at which the resolution changes, then a new approximation consisting of L^d points, where $L \approx M/r$, must be computed. First the values of these numbers are taken as multilinear or higher order interpolation of the M^d old values. (2) A random element of variance proportional to $1/(L-1)^{2H}$ is added to all points. These two steps are repeated until we have obtained N^d points. For details see [Saup88].

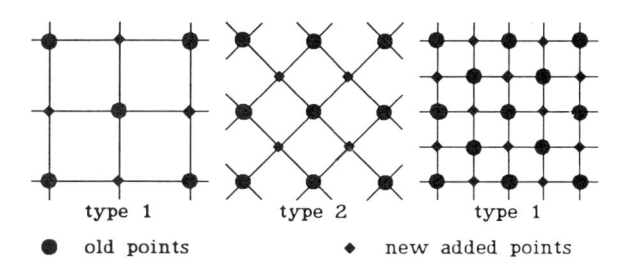

Figure 8. Grid types for displacement methods. In the displacement methods a grid of type 1 is given at the beginning, from which a type 2 grid is generated. Its mesh size is $1/\sqrt{2}$ times the old mesh size. In a similar step a grid of type 1 is again obtained, as shown in the figure.

ALGORITHM Title	\multicolumn{2}{l	}{MidPointFM2D(X, maxlevel, sigma, H, addition, seed) Midpoint displacement and successive random additions in two dimensions}
Arguments	X[] []	doubly indexed real array of size $(N+1)^2$
	maxlevel	maximal number of recursions, $N = 2^{maxlevel}$
	sigma	initial standard deviation
	H	parameter H determines fractal dimension $D = 3 - H$
	addition	boolean parameter (turns random additions on/off)
	seed	seed value for random number generator
Variables	i, N, stage	integers
	delta	real holding standard deviation
	x, y, y0, D, d	integer array indexing variables
Functions	\multicolumn{2}{l	}{f3(delta,x0,x1,x2) = (x0+x1+x2)/3 + delta*Gauss()}
	\multicolumn{2}{l	}{f4(delta,x0,x1,x2,x3) = (x0+x1+x2+x3)/4 + delta*Gauss()}

BEGIN
   InitGauss(seed)
   N = power(2, maxlevel)
                    /* set the initial random corners */
   delta := sigma
   X[0][0] := delta*Gauss()
   X[0][N] := delta*Gauss()
   X[N][0] := delta*Gauss()
   X[N][N] := delta*Gauss()
   D := N
   d := N / 2

   FOR stage := 1 TO maxlevel DO
                    /* going from grid type I to type II */
      delta := delta*power(0.5, 0.5*H)
                    /* interpolate and offset points */
      FOR x := d TO N-d STEP D DO
         FOR y := d TO N-d STEP D DO
            X[x][y] := f4 (delta, X[x+d][y+d], X[x+d][y-d], X[x-d][y+d]
                 X[x-d][y-d])
         END FOR
      END FOR
                    /* displace other points also if needed */

      IF (addition) THEN
         FOR x := 0 TO N STEP D DO
            FOR y := 0 TO N STEP D DO
               X[x][y] := X[x][y] + delta*Gauss()
           END FOR
         END FOR
      END IF
   (continued on the next page)

```
ALGORITHM    MidPointFM2D(X, maxlevel, sigma, H, addition, seed)
Title        Midpoint displacement and successive random additions
             in two dimensions (continued from previous page)
```

```
                /* going from grid type II to type I */
        delta := delta*power(0.5, 0.5*H)
                /* interpolate and offset boundary grid points */
        FOR x := d TO N-d STEP D DO
            X[x][0] := f3(delta, X[x+d][0], X[x-d][0], X[x][d])
            X[x][N] := f3(delta, X[x+d][N], X[x-d][N], X[x][N-d])
            X[0][x] := f3(delta, X[0][x+d], X[0][x-d], X[d][x])
            X[N][x] := f3(delta, X[N][x+d], X[N][x-d], X[N-d][x])
        END FOR
                /* interpolate and offset interior grid points */
        FOR x := d TO N-d STEP D DO
            FOR y := D TO N-d STEP D DO
                X[x][y] := f4(delta, X[x][y+d], X[x][y-d], X[x+d][y], X[x-d][y])
            END FOR
        END FOR
        FOR x := D TO N-d STEP D DO
            FOR y := d TO N-d STEP D DO
                X[x][y] := f4(delta, X[x][y+d], X[x][y-d], X[x+d][y], X[x-d][y])
            END FOR
        END FOR
                /* displace other points also if needed */
        IF (addition) THEN
            FOR x := 0 TO N STEP D DO
                FOR y := 0 TO N STEP D DO
                    X[x][y] := X[x][y] + delta*Gauss()
                END FOR
            END FOR
            FOR x := d TO N-d STEP D DO
                FOR y := d TO N-d STEP D DO
                    X[x][y] := X[x][y] + delta*Gauss()
                END FOR
            END FOR
        END IF
        D := D/2
        d := d/2
    END FOR
END
```

THE FOURIER FILTERING METHOD

We now proceed to the extension of the Fourier filtering method to higher dimensions. In two dimensions the spectral density S generally will depend on two

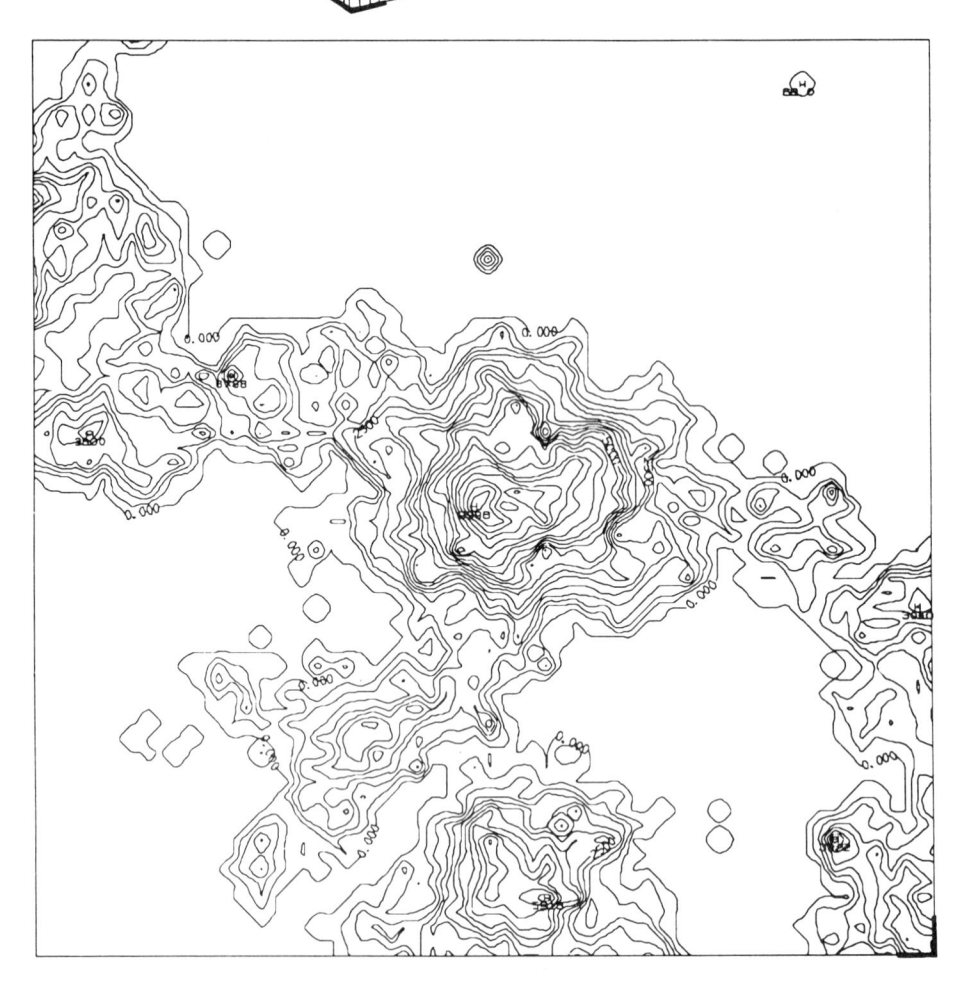

Figure 9. Topographical map and perspective view of random fractal ($H = 0.8$).

ALGORITHM	InterpolatedFM1D(X, N, r, sigma, H, seed)
Title	One-dimensional fractal motion with lacunarity

Arguments	X[ ]	real array of size N
	N	number of elements in X
	r	scaling factor for resolutions $(0 < r < 1)$
	sigma	initial standard deviation
	H	$0 < H < 1$ determines fractal dimension $D = 2 - H$
	seed	seed value for random number generator
Variables	delta	real variable holding standard deviations Δ
	Y[ ]	real array of size N for the interpolated values of X
	mT,mt	integers, number of elements in arrays X and Y
	t,T	sampling rates for X and Y
	i,index	integer
	h	real

```
BEGIN          /* initialize the array with 2 points */
    InitGauss(seed)
    X[0] := 0
    X[1] := sigma*Gauss()
    mT := 2
    T := 1.0
                /* loop while less than N points in array */
    WHILE (mT < N) DO
                /* set up new resolution of mt points */
        mt := mT/r
        IF (mt = mT) THEN mt := mT + 1
        IF (mt > N) THEN mt := N
        t := 1/(mt - 1)
                /* interpolate new points from old points */
        Y[0] := X[0]
        Y[mt - 1] := X[mT - 1]
        FOR i := 1 TO mt - 2 DO
            index := integer(i*t/T)
            h := i*t/T - index
            Y[i] := (1 - h)*X[index] + h*X[index + 1]
        END FOR
                /* compute the standard deviation for offsets */
```
$$\text{delta} := \text{sqrt}(0.5)*\text{power}(t,H)*\text{sigma}*\text{sqrt}\Big(1.0 - \text{power}(t/T, 2 - 2*H)\Big)$$
```
                /* do displacement at all positions */
        FOR i := 0 TO mt - 1
            X[i] := Y[i] + delta*Gauss()
        END FOR
        mT := mt
        T := 1/mT
    END WHILE
END
```

frequency variables u and v, corresponding to the x and y directions. But since all directions in the xy plane are equivalent with respect to statistical properties, S depends only on $\sqrt{u^2 + v^2}$. If we cut the surface along a straight line in the xy plane, we expect for the spectral density S of this fBm in only one dimension to be a power law $1/f^\beta$, as before. This requirement implies (see Eq. 16) that the two-dimensional spectral density will behave like

$$S(u,v) = \frac{1}{(u^2+v^2)^{H+1}}$$

The two-dimensional discrete inverse Fourier transform is

$$X(x,y) = \sum_{k=0}^{N-1}\sum_{l=0}^{N-1} a_{kl} e^{2\pi i(kx+ly)}$$

for $x,y = 0, 1/N, 2/N, \cdots, (N-1)/N$, and thus we specify for the coefficients a_{kl}

$$E(|a_{kl}|^2) \propto \frac{1}{(k^2+l^2)^{H+1}}$$

Since the constructed function X is a real function, we must also satisfy a conjugate symmetry condition, namely

$$a_{N-i,N-j} = \overline{a_{i,j}} \text{ for } i,j > 0$$
$$a_{0,N-j} = \overline{a_{0,j}} \text{ for } j > 0$$
$$a_{N-i,0} = \overline{a_{i,0}} \text{ for } i > 0$$
$$a_{0,0} = \overline{a_{0,0}}$$

The fractal dimension D_f of the surface will be $D_f = 3 - H$. The algorithm then consists simply of choosing the Fourier coefficients accordingly and then performing a two-dimensional inverse transform.

In the sequence in Figure 10 we show how the spectral synthesis method 'adds detail' by improving the spectral representation of the random fractal, i.e., by allowing more and more Fourier coefficients to be employed. The resolution in the algorithm was $N = 64$, but in the top image of Figure 10 only the first $2^2 = 4$ coefficients were used. In the other pictures we allowed $8^2 = 64$ (middle) and $32^2 = 1024$ (bottom) nonzero coefficients.

Point Evaluation of Multivariable Random Fractals

A technique very much related to random fractals was introduced by Peachey and Perlin in 1985 [Peac85; Perl85]. In this framework one may consider random functions of three variables as models of *solid texture*. The functions are generated procedurally, and the approach has been termed *functional-based modelling*.

Random Fractals in Image Synthesis 111

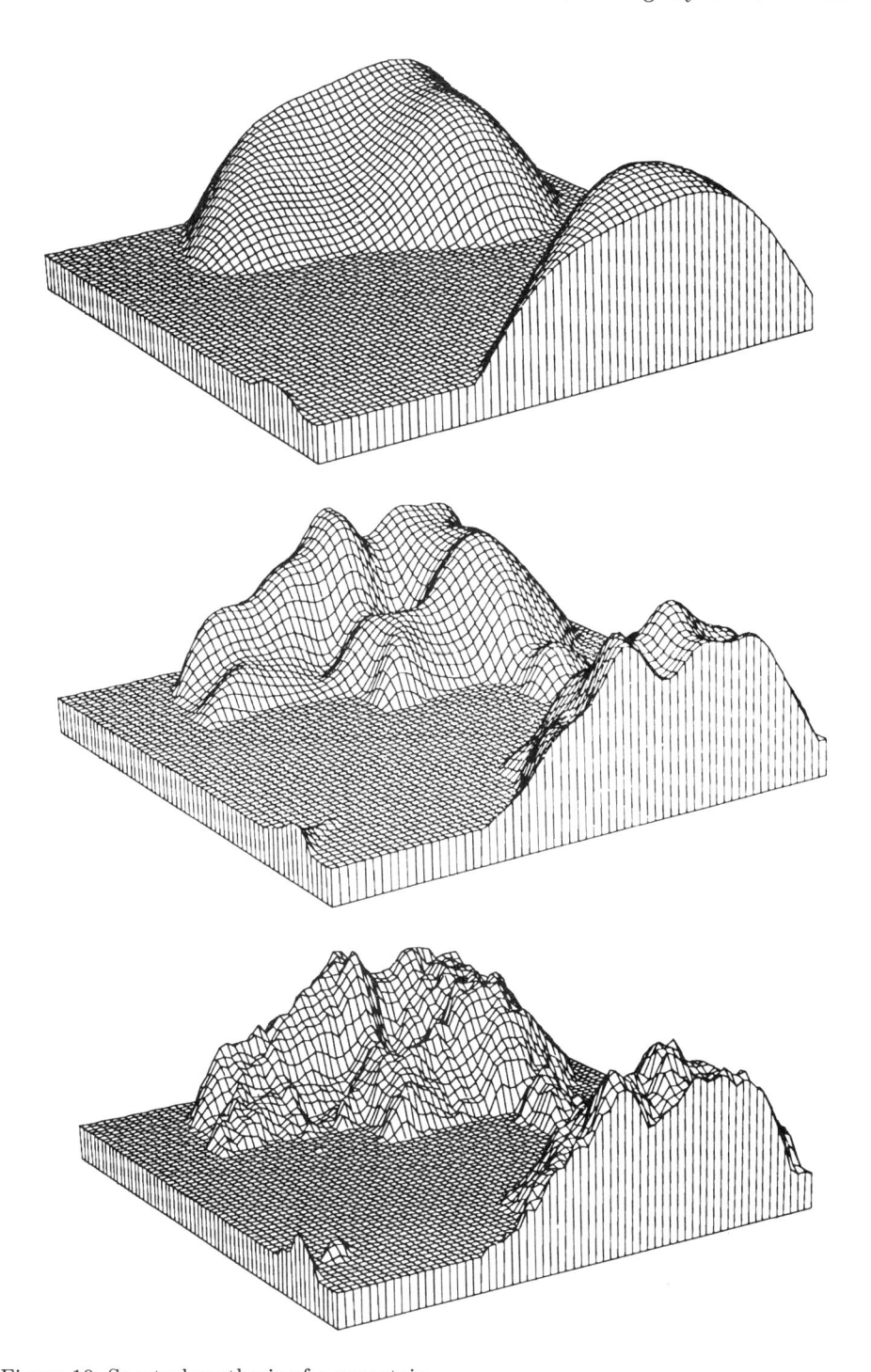

Figure 10. Spectral synthesis of a mountain.

In the following we suggest blending Perlin's techniques with others stemming from random fractals, in particular the Mandelbrot-Weierstrass function. As a result we obtain a simple and fast method which has some striking advantages over the other algorithms for the generation of random fractals.

- The fractal dimension may be variable, e.g., we can let the dimension D be a function $D = D(x_1, \cdots, x_n)$ varying in space.
- Although a random fractal function at a given point generally depends on its values at all other points, the calculation is a point evaluation, and no explicit references to other points are necessary. Therefore the order of the computation does not matter, which is important for implementations on parallel processors.
- The method is relatively simple to implement and still offers a significant increase in speed for many cases of interest.

The goal of the construction is a random function $X_3(x, y, z)$ of three variables which take real values. All points (x, y, z) and all directions should be statistically equivalent. Moreover, in the frequency domain the spectral density function is of the form $1/f^\beta$, where f denotes the combined spatial frequencies in the x, y and z directions, i.e., $f = \sqrt{f_x^2 + f_y^2 + f_z^2}$.

Let us begin with a one-dimensional model $X_1(x)$. First an auxiliary function $S_1(x)$ is defined at integer points $k \in \mathbb{Z}$. The value $S_1(k)$ is taken as a sample of a random variable with mean zero (e.g., a Gaussian random variable). Then a differentiable, interpolating function, $S_1(x)$, is constructed from the data. Here one can choose piecewise cubic Hermite interpolation with the additional constraints $S_1'(k) = 0$ for integers k (see [Peac88] and below for the 3D case). Other choices are possible, of course. In Perlin [Perl85] the auxiliary function S_1 is called 'noise'. But S_1 contains primarily low frequencies and therefore should not be conceived as $1/f$ noise. However, as we sum up properly scaled-down copies of S_1 we obtain a $1/f$ noise. More precisely, the one-dimensional noise function is defined as

$$X_1(x) = \sum_{k=0}^{\infty} \frac{1}{r^{kH}} S_1(r^k x) \tag{18}$$

with $r > 1, 0 \leq H \leq 1$. This is very similar to the original *turbulence* function used in [Perl85], i.e., $\sum_{k=0}^{\infty} 2^{-k} |S_1(2^k x)|$. As k grows the contributions to the sum rapidly get small. Thus, in practice the summation is carried out over only a few terms (see Figure 11).

The generalization to two, three or more dimensions is straightforward. For example, in the two-dimensional case we define an auxiliary function $S_2(x, y)$ first at integer lattice points (k, l) by again sampling a random variable. Secondly, the data is interpolated by a smooth function. The noise function becomes

$$X_2(x, y) = \sum_{k=0}^{\infty} \frac{1}{r^{kH}} S_2(r^k x, r^k y) \tag{19}$$

Clearly, $X_2(x,y)$ is a continuous function of x and y. However, we point out that not all directions in the xy plane are equivalent. On the x- and y-axes one random data point is entered per unit interval, whereas, e.g., on the diagonal the distance between random data points is $\sqrt{2}$. This necessarily shrinks the width of the power spectral density of $S_2(x,y)$: The power spectrum of $X_2(x,y)$ decays by a factor of about $\sqrt{2}$ faster along the diagonal. In all of our experiments this lack of purity has not been visually noticeable.

In the three-dimensional case, which is the most interesting case in applications, we thus define the noise function $X_3(x,y,z)$

$$X_3(x,y,z) = \sum_{k=0}^{\infty} \frac{1}{r^{kH}} S_3(r^k x, r^k y, r^k z) \qquad (20)$$

The interpolation as suggested by Peachey [Peac88] and Perlin [Perl85] is as follows: For a given point (x, y, z) let us write

$$x = i_x + d_x, \qquad y = i_y + d_y, \qquad z = i_z + d_z$$

where i_x, i_y, i_z are integers and d_x, d_y, d_z are nonnegative fractions less than 1. Then set

$$s_x = d_x^2(3 - 2d_x), \qquad s_y = d_y^2(3 - 2d_y), \qquad s_z = d_z^2(3 - 2d_z)$$

and define

$$\begin{aligned}S_3(x,y,z) =\,& s_x s_y s_z S_3(i_x+1, i_y+1, i_z+1) \\&+ (1-s_x) s_y s_z S_3(i_x, i_y+1, i_z+1) \\&+ s_x(1-s_y) s_z S_3(i_x+1, i_y, i_z+1) \\&+ (1-s_x)(1-s_y) s_z S_3(i_x, i_y, i_z+1) \\&+ s_x s_y (1-s_z) S_3(i_x+1, i_y+1, i_z) \\&+ (1-s_x) s_y (1-s_z) S_3(i_x, i_y+1, i_z) \\&+ s_x(1-s_y)(1-s_z) S_3(i_x+1, i_y, i_z) \\&+ (1-s_x)(1-s_y)(1-s_z) S_3(i_x, i_y, i_z) \end{aligned} \qquad (21)$$

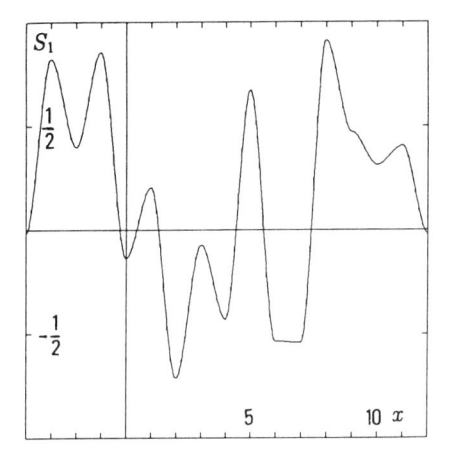

Figure 11. A sample of an auxiliary function $S_1(x)$.

This piecewise cubic interpolation yields continuously differentiable real function in $\mathbf{R}^3$. It is clear how the procedure can be extended to produce random fractal functions in four, five and more variables (see Figure 12). In summary we have:

Rescale-and-add method: Let $S_n : \mathbf{R}^n \to \mathbf{R}$ be a real function that has values at integer lattice points of $\mathbf{R}^n$ defined by Gaussian random variables of zero mean and the same variance at all points. Further let $S_n(x_1, ..., x_n)$ be a smooth interpolation from the data at the integer lattice points. Then the function $X_n : \mathbf{R}^n \to \mathbf{R}$ defined by

$$X_n(x) = \sum_{k=k_0}^{\infty} \frac{1}{r^{kH}} S_n(r^k x) \qquad (22)$$

with $r > 1$, $0 \leq H \leq 1$ and $k_0 \leq 0$ is a random function whose graph has a fractal dimension

$$D = n + 1 - H \qquad (23)$$

and $r > 1$ determines lacunarity.

The implementation of Eq. (22) is not difficult. One of the two issues that must be considered is the range of the summation. The lower index k_0 should obviously be chosen small enough so that the largest scale L of the objects is not greater than r^{-k_0}, i.e.,

$$r^{k_0} L \leq 1 \qquad (24)$$

If k_0 is too big one sees the effect of the dominant frequencies (≈ 1) of the auxiliary function S_n. The cutoff at the other end of the summation is harder to define. If the random function will be sampled at points which have a distance Δ from each other, then we must be aware of aliasing effects if the summation is carried out with terms exceeding the Nyquist limit, i.e., terms with $r^k > 2/\Delta$. This motivates a cutoff at

$$k \approx \frac{\log 2/\Delta}{\log r} \qquad (25)$$

The corresponding discussion of Peachey [Peac88] also applies here. However, if H is small then the amplitudes of the clamped terms may still be very large, and an aesthetically more pleasing picture may be obtained if the summation in Eq. (22) is continued for a few more terms than suggested by Eq. (25).

The other remaining problem is the production of the random numbers attached to the integer lattice points. We suggest using a three-dimensional array T

$$T[i, j, k], \qquad i, j, k = 0, \cdots, N - 1$$

For given integers x, y, z we define integers i_x, i_y, i_z as in

$$i_x = \begin{cases} x \bmod N & \text{if } x \geq 0 \\ x \bmod N + N & \text{if } x < 0 \end{cases}$$

Then we let $X(x, y, z) = T[i_x, i_y, i_k]$.

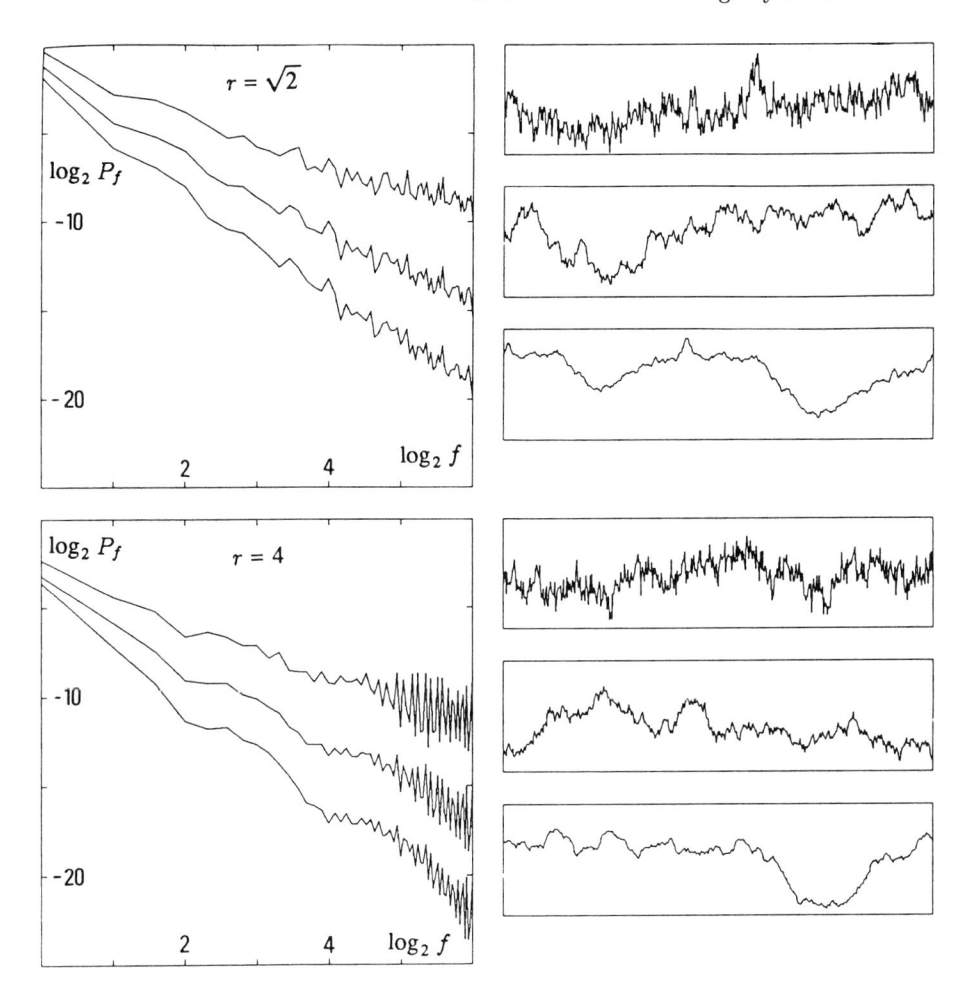

Figure 12. Power spectra and noise samples for different parameters of the rescale-and-add method. The lacunarity parameter r in the top graphs is $r = \sqrt{2}$, in the bottom graphs $r = 4$. Curves for three values of H are shown: $H = 0.2$ (top), $H = 0.5$ (center) and $H = 0.8$ (bottom). The corresponding exponent β of the $1/f^\beta$ power law is $\beta = 1 + 2H$, i.e., $\beta = 1.4$ (top), $\beta = 2.0$ (center) and $\beta = 2.6$ (bottom). These are actually the approximate negative slopes of the empirically computed power spectra, as shown. On the right, corresponding random fractal functions (samples of $X_1(x)$) are given. The spectral density plots are obtained from averaging squared amplitudes of several periodograms of samples of $X_1(x)$.

The rescale-and-add method may be modified for a variety of effects, such as the modulation of amplitudes and fractal dimension by height and by external functions (see Plates 22 and 23). Following Musgrave et al. [Musg89a], modulation by height can be accomplished as follows: let

$$X_n(x) = \lim_{k \to \infty} Y_k(x)$$

where
$$Y_0(x) = S_n(x)$$
$$Y_{k+1}(x) = Y_k(x) + \frac{S_n(r^{kH_k(Y_k(x))}x)}{r^{kH_k(Y_k(x))}}$$

and $H_k(Y_k(x))$ is a function of the so far accumulated 'height' $Y_k(x)$. In this fashion one can produce ragged mountain tops together with smooth valleys.

Another application is the simulation and animation of clouds [Saup89b], where the underlying basis is the observation that clouds are fractal not only in space but also in time [Love85]. Thus, random fractals with three or even four variables are applicable. One example is given in Plate 24. The values of the random fractal function are interpreted as water vapor densities D. These values are mapped via a color ramp ranging from blue for low density to white and dark grey for higher densities. For simplicity clouds can be taken as two-dimensional, with (screen) coordinates u and v. An animation of changing clouds with some effect of wind might most easily be modelled by

$$D(u,v,t) = X(x(u,v,t), y(u,v,t), z(u,v,t))$$

where X denotes a random fractal function with three variables, and t denotes time. In a first attempt, the dependencies of x, y, z on u, v, t may be set to

$$x(u,v,t) = u + td_x$$
$$y(u,v,t) = v + td_y$$
$$z(u,v,t) = td_z$$

where d_x, d_y and d_z denote some constants which determine the speed at which the scene changes. Plate 24 shows a series of frames of a cloud development with some added 'turbulence' effect.

Acknowledgements. The author thanks R.F. Voss for sharing his expertise on random fractals and F.K. Musgrave for his kind permission to reproduce two of his color images here. I. Nikschat-Tillwick and O. Sachse have written the programs to render the planet in Plate 25.

REFERENCES

[Fell85]
    Fellous, A., Granada, J., and Hourcade, J., Fractional Brownian relief: An exact local method, in *Proc. Eurographics '85*, New York: North Holland, pp. 353–363, 1985.

[Four82]
    Fournier, A., Fussell, D., and Carpenter, L., Computer rendering of stochastic models, *CACM*, Vol. 25, pp. 371–384, 1982.

[Four85]
Fournier, A. and Milligan, T., Frame buffer algorithms for stochastic models, *IEEE Comput. Graph. and Appl.*, Vol. 5, p. 10, 1985.

[Harr87]
Harrington, S., *Computer Graphics—A Programming Approach*, New York: McGraw Hill, 1987.

[Hear86]
Hearn, D. and Baker, M.P., *Computer Graphics*, Englewood Cliffs, NJ: Prentice-Hall, 1986.

[Lewi87]
Lewis, J.P., Generalized stochastic subdivision, *ACM Trans. on Graph.*, Vol. 6, No. 3, pp. 167–190, 1987.

[Lewi89]
Lewis, J.P., Algorithms for solid noise synthesis, *Comput. Graph.*, Vol. 23, pp. 263–270, 1989 (SIGGRAPH 89).

[Love85]
Lovejoy, S., and Mandelbrot, B.B., Fractal properties of rain, and a fractal model, *Tellus*, Vol. 37A, pp. 209–232, 1985.

[Mand68]
Mandelbrot, B.B., and Van Ness, J.W., Fractional Brownian motion, fractional noises and applications, *SIAM Review*, Vol. 10, pp. 422–437, 1968.

[Mand82a]
Mandelbrot, B.B., *The Fractal Geometry of Nature*, New York: W.H. Freeman, 1982.

[Mand82b]
Mandelbrot, B.B., Comment on computer rendering of fractal stochastic models, *CACM*, Vol. 25, pp. 581–583, 1982.

[Mand88]
Mandelbrot, B.B., Fractal Landscapes without Creases and with Rivers, in *The Science of Fractal Images*, Peitgen, H.-O., and Saupe, D., Eds., New York: Springer-Verlag, 1988.

[Mill86]
Miller, G., The definition and rendering of terrain maps, *Comput. Graph.*, Vol. 20, pp. 39–48, 1986.

[Musg89a]
Musgrave, F.K., Kolb, C., and Mace, R., The synthesis and rendering of eroded fractal terrains, *Comput. Graph.*, Vol. 23, pp. 41–50, 1989 (SIGGRAPH 89).

[Musg89b]
Musgrave, F.K., Prisms and rainbows: A dispersion model for computer graphics, *Proc. Graphics Interface '89—Vision Interface '89*, London, Ontario, Canada, June 1989.

[Peac85]
Peachey, D.R., Solid texturing of complex surfaces, *Comput. Graph.*, Vol. 19, pp. 279–286, 1985 (SIGGRAPH 85).

[Peac88]
Peachey, D.R., Solid textures and anti-aliasing issues, in Functional-based modelling, SIGGRAPH 88 Course Notes 28, Atlanta, 1988.

[Peit88]
Peitgen, H.-O., and Saupe, D., Eds., *The Science of Fractal Images*, New York: Springer-Verlag, 1988.

[Perl85]
Perlin, K., An image synthesizer, *Comput. Graph.*, Vol. 19, pp. 287–296, 1985 (SIGGRAPH 85).

[Rume90]
Rümelin, W., Simulation of fractional Brownian motion, in *Fractal 90—Proc. First IFIP Conf. Fractals*, (Lisbon, June 6–8, 1990), Peitgen, H-O., Henriques, J.M., and Penedo, L.F., Eds., Amsterdam: Elsevier, 1990.

[Saup88]
Saupe, D., Algorithms for Random Fractals, in *The Science of Fractal Images*, Peitgen, H.-O., and Saupe, D., Eds., New York: Springer-Verlag, 1988.

[Saup89a]
Saupe, D., Point Evaluation of Multi-variable Random Fractals, in *Visualisierung in Mathematik und Naturwissenschaften*, Bremer Computergrafiktage 1989, Jürgens, H., and Saupe, D., Eds., Heidelberg: Springer-Verlag, pp. 114–126, 1989.

[Saup89b]
Saupe, D., Simulation und Animation von Wolken mit Fraktalen, in *19. GI-Jahrestagung I*, Paul, M., Ed., Heidelberg: Springer-Verlag, pp. 373–384, 1989.

[Voss85]
Voss, R.F., Random Fractal Forgeries, in *Fundamental Algorithms for Computer Graphics*, Earnshaw, R.A., Ed., Berlin: Springer-Verlag, pp. 805–835, 1985.

IFSs and the Interactive Design of Tiling Structures

Alastair N. Horn

Abstract

Deterministic fractal geometry provides a framework for describing both the geometry of man-made structures and the geometry of nature.

A large class of deterministic fractals are those which may be 'partitioned' into a number of 'tiles'. We term such fractals tiling structures. We describe a representation scheme for tiling structures called Iterated Function Systems (IFSs). IFSs make explicit the mappings which take a tiling structure into its tiles.

We show how tiling structures exist as the limit of both random and deterministic processes based upon the IFS, and we exploit massive SIMD parallelism in the generation of tiling structures and in the rendering of their images.

We also attempt to answer the question, Can one synthesise tiling structures by the interactive graphical manipulation of a representation of IFSs? and present our Interactive System for Image Synthesis (ISIS).

Introduction

In making sense of the universe about us we invent geometries which make our spatial intuitions objective. Mandelbrot suggested the existence of geometries near to the 'geometry of nature' [Mand82]. Mandelbrot's fractal geometry [Falc85] provides a framework for the description of fantastic structures of apparently infinite complexity [Peit86]. We are fortunate that in many cases fractals have simple descriptions and exist as the limit of simple iterative processes.

A large class of deterministic fractals are those which may be 'partitioned' into a number of 'tiles'. We term such fractals *tiling structures*. Tiling structures may be represented by a scheme called *Iterated Function Systems* (IFSs). IFSs were hinted at in work by Hutchinson [Hutc81] on self-similar sets and later named, formalised and popularised in important work by Barnsley and Demko [Barn85] and others.

In this paper we introduce tiling structures and show how they exist as the limit of both random and deterministic processes currently exploited in computer graphics [Peit88]. We describe how IFSs provide both a representation for tiling structures and a basis for both random and deterministic processes which generate tiling structures. We show how massive *Single Instruction Multiple Data* (SIMD) parallelism can be exploited in the generation of tiling structures and in the rendering of their images. Finally, we attempt to answer the question, Can one synthesise tiling structures by the interactive graphical manipulation of a representation of IFSs? and we present our *Interactive System for Image Synthesis* (ISIS).

Tiling Structures

A 'tiling structure' is a structure which may be 'partitioned' into a number of 'tiles' [Grun87]. Building on this loose definition, a taxonomy of tiling structures may be suggested consisting of two main classes [Horn90]:

> A tiling structure which is composed of tiles equivalent to the structure under some mapping is said to be *self-tiling*. For example, a square may be considered self-tiling.

> A tiling structure which is composed of tiles equivalent to the structure or to external structures under some mapping is said to be *metatiling*. For example, a row of houses vanishing to infinity may be considered metatiling (how, we will see later).

Structures may exhibit tiling properties in a strict sense, or only over a limited range of spatial scales, or only in a statistical sense. The domain of tiling structures is rich in structures which are reminiscent of both man-made and natural structures.

FRACTAL TILING STRUCTURES IN COMPUTER GRAPHICS

In the field of computer graphics fractal tiling structures are much in evidence. Both the complex structures based upon statistical fractal geometry, and the intricate structures which are the outcome of deterministic rewriting systems, are in some sense tiling structures.

Statistical fractal geometry describes structures by making correlations of their expected characteristics. A common primitive of statistical fractal geometry is fractional Brownian motion (fBm) [Mand68]. Fournier and others [Four82] have used a midpoint displacement algorithm (invented 2000 years earlier by Archimedes in his construction of the parabola) based upon a triangular-subdivision scheme which approximates fBm. They have applied it to modelling mountain ranges (Figure 1) and rough terrain: Starting with a single triangle, the midpoints of the sides of each triangle are joined to form four quarter-size triangles. At each level of resolution a suitably scaled random variable is used

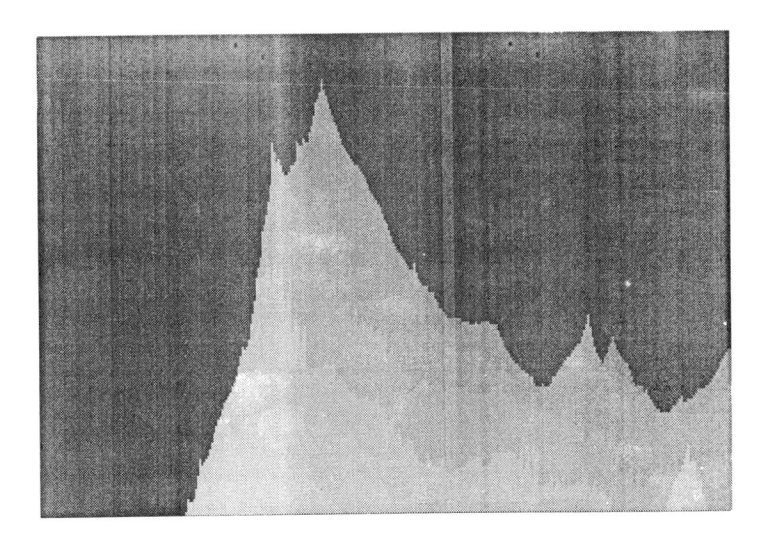

Figure 1. fBm mountain-scape (after Fournier et al. [Four82]).

to perturb a height field at each vertex created (or all vertices in a variant of the method). These structures exhibit statistical tiling properties related to the scaling properties of fBm.

Fractals may also be expressed as rewriting systems (also called grammars, or L-systems) [Mand82]. The simplest rewriting systems are of the context-free kind, which are based upon a single base element and a generator (see Figure 2). Starting with a single instance of the base element, a unique structure is defined solely by the iterative application of a rewriting rule. The rewriting rule is described by the generator. Each base element is replaced by one or more instances of the base element identical under some mapping.[†] Rewriting systems are simply a notation for describing tiling structures. Providing the rewriting rule is deterministic, the rewriting system defines a unique tiling structure. The tiling structure is defined by the mappings implicit in the rewriting rule.

The structures and images produced by both fBm and rewriting systems are visually stunning, but their production is at present more an art than a science. The abundance of fractal tiling structures in computer graphics has only recently been noticed, through a new application of deterministic fractal geometry which we now describe.

[†]Rewriting systems typically take the base element to be a vector and the generator to be a connected set of vectors describing a path in space. This imposes restrictions on the variety of structures that can be represented by a rewriting system using only a single base element. This limitation is often avoided by having several different base elements, each with their own rewriting rules, with the tradeoff of additional complexity in the structure representation. A better choice of rewriting system may often be found if the generator is not constrained to describe a path.

IFSs

We regard a structure as being a compact subset of some complete metric space [Suth75]. A structure can be thought of as a distribution of particles in some space, e.g., ink on a page. A (higher) space may also be defined whose points correspond to structures.

Definition 1 *Let* $(\mathcal{X}, d)$ *be a complete metric space, with metric d. And let* $(\mathcal{H}(\mathcal{X}), h(d))$ *denote the corresponding space of nonempty compact subsets, or structures, of* $(\mathcal{X}, d)$, *with metric* $h(d)$.

We are all familiar with mappings (or transformations) of one sort or another. Contraction mappings bring points closer together.

Definition 2 (Contraction Mapping) *Let* $w : \mathcal{X} \to \mathcal{X}$ *be a mapping on* $(\mathcal{X}, d)$. *Then we define the contractivity (or Lipschitz constant) of w by*

$$Lip\ w = max_{x \neq y} \frac{d(w(x)), (w(y))}{d(x, y)} \quad for\ all\ x, y \in \mathcal{X}$$

If $Lip\ w = s$, *then* $d(w(x), w(y)) \leq sd(x, y)$, *and if* $s < 1$ *we say w is a contraction mapping.*

We now describe features of a representation scheme for tiling structures called Iterated Function Systems (IFSs).

THE SIMPLE IFS

Self-tiling structures are defined in terms of contraction mappings in a simple and flexible way. A self-tiling structure is represented by a simple IFS; the explicit connection between the IFS and the tiling we will bring out later.

Definition 3 (Simple IFS) *A simple IFS on* $(\mathcal{X}, d)$ *is an ordered set of contraction mappings on* $(\mathcal{X}, d)$

$$\{w_j : j = 1, \ldots, n\}$$

The contractivity of the simple IFS is defined to be the greatest Lipschitz constant of the ordered set of mappings.

We can promote a contraction mapping on $(\mathcal{X}, d)$ to a contraction mapping on the space $(\mathcal{H}(\mathcal{X}), h(d))$, i.e., a mapping which takes a structure into a structure rather than a point into a point.

Definition 4 (Mapping Promotion) *Let* $w : \mathcal{X} \to \mathcal{X}$ *be a mapping on* $(\mathcal{X}, d)$ *with contractivity s. Then* $w : \mathcal{H}(\mathcal{X}) \to \mathcal{H}(\mathcal{X})$ *is defined by*

$$w(B) = \{w(x) : x \in B\} \quad for\ all\ B \in \mathcal{H}(\mathcal{X})$$

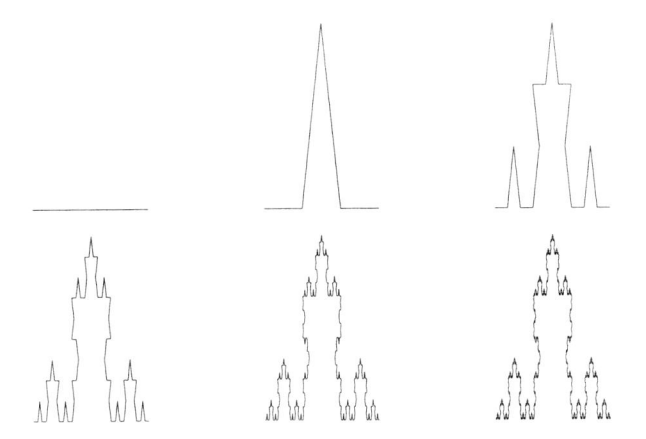

Figure 2. Koch curve rewriting system: The base element is a vector; the generator is made up of four mappings of the base element.

A simple IFS defines a mapping which takes a structure into a structure.

Definition 5 (Simple IFS Mapping) *Let $\{w_j : j = 1, \ldots, n\}$ be a simple IFS on $(\mathcal{X}, d)$. Then the mapping of the simple IFS $W : \mathcal{H}(\mathcal{X}) \to \mathcal{H}(\mathcal{X})$ is defined by*

$$W(B) = w_1(B) \cup w_2(B) \cup \ldots \cup w_n(B)$$

$$= \bigcup_{j=1}^{j=n} w_j(B) \; for \; all \; B \in \mathcal{H}(\mathcal{X})$$

In the case that $h(d)$ is the Hausdorff metric, Hutchinson [Hutc81] has proved that $w(B)$ is a contraction mapping on $(\mathcal{H}(\mathcal{X}), h(d))$, with equal contractivity to $w(x)$, and that $W(B)$ is a contraction mapping on $(\mathcal{H}(\mathcal{X}), h(d))$, with contractivity equal to that of the simple IFS. These results hold in other metrics.

Banach's fixed point theorem states that every contraction mapping in a complete metric space has a unique fixed point. The fixed point of the mapping of a simple IFS is a unique compact subset of $(\mathcal{X}, d)$ (a self-tiling structure).

Lemma 1 (Attractor) *Let $\{w_j : j = 1, \ldots, n\}$ be a simple IFS on $(\mathcal{X}, d)$, with mapping W. Its unique fixed point, $A \in (\mathcal{H}(\mathcal{X}), h(d))$, obeys*

$$A = W(A) = \bigcup_{j=1}^{j=n} w_j(A)$$

and is called the attractor of the simple IFS.

The attractor of a simple IFS is a self-tiling structure. The attractor A is covered by n tiles, which are mappings of itself. An example of a self-tiling structure is the square. The square is tiled by its four quadrants, hence its IFS is simply the four affine mappings taking the square into its quadrants. Note that this is not the only possible IFS representation for a square (see [Horn90]).

THE COLLAGE THEOREM

The notion of tiling is made explicit by an important theorem due to Barnsley and Demko [Barn85]. The Collage Theorem quantifies the nearness of the attractor of a simple IFS to any given structure, the mappings of the IFS being chosen to define an approximate self-tiling of the structure. The Collage Theorem tells us that we need only deduce an approximate self-tiling of any structure in order to approximately describe it using a simple IFS (see Figure 3).

Theorem 1 (Collage Theorem) *Let $L \in \mathcal{H}(\mathcal{X})$, and let*

$$\{w_j : j = 1, \ldots, n\}$$

be a simple IFS on $(\mathcal{H}(\mathcal{X}), h(d))$ with contractivity s, such that

$$h\left(L, \bigcup_{j=1}^{j=n} w_j(L)\right) \leq \epsilon$$

Then

$$h(L, A) \leq \frac{\epsilon}{(1-s)}$$

where A is the attractor of the simple IFS.

THE META-IFS

Although the domain of self-tiling structures is infinitely large, *metatiling* structures widen the scope of IFSs to include tiling structures which are not the same at all scales. Metatiling structures are described by meta-IFSs [Horn90]. We associate an order with a meta-IFS, a zeroth order meta-IFS being a simple IFS.

Definition 6 (Meta-IFS) *An order-m meta-IFS on $(\mathcal{X}, d)$ is a simple IFS on $(\mathcal{X}, d)$*

$$\{w_j : j = 1, \ldots, n\}$$

and an ordered set of points in $(\mathcal{H}(\mathcal{X}), h(d))$ (external structures)

$$\{v_1, v_2, \ldots, v_m\}$$

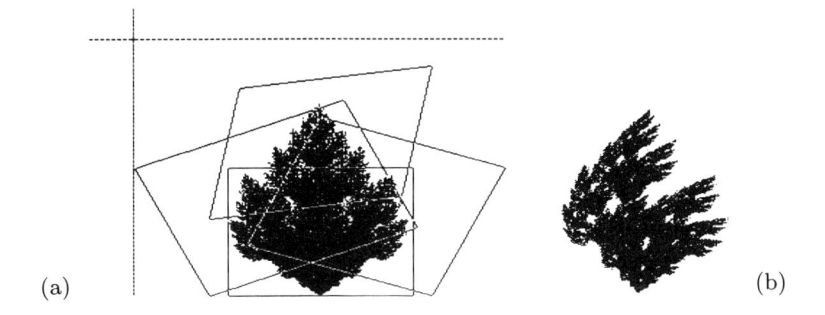

Figure 3. Collage Theorem. (a) An approximate tiling of a leaf; (b) the attractor.

An order-m meta-IFS is denoted by

$$\{\{w_j : j = 1, \ldots, n\}; \quad v_i : i = 1, \ldots, m\}$$

The contractivity of the meta-IFS is defined to be the greatest Lipschitz constant of the ordered set of mappings.

Note that a point $v \in (\mathcal{H}(\mathcal{X}), h(d))$ may be considered a mapping with zero contractivity, which takes every point in $(\mathcal{H}(\mathcal{X}), h(d))$ to the point v. In this fashion, a meta-IFS defines a mapping which takes a structure into a structure.

Lemma 2 (Meta-IFS Mapping) *Let* $\{\{w_j : j = 1, \ldots, n\}; v_i : i = 1, \ldots, m\}$ *be a meta-IFS on* $(\mathcal{X}, d)$. *Then the mapping of a meta-IFS* $W : \mathcal{H}(\mathcal{X}) \to \mathcal{H}(\mathcal{X})$ *is defined by*

$$W(B) = w_1(B) \cup \ldots \cup w_n(B) \cup v_1 \cup \ldots \cup v_m$$
$$= \bigcup_{j=1}^{j=n} w_j(B) \cup \bigcup_{i=1}^{i=m} v_i$$

The mapping of a meta-IFS is again a contraction mapping; its fixed point is a unique compact subset of $(\mathcal{X}, d)$ (a metatiling structure).

Lemma 3 (Attractor) *Let* $\{\{w_j : j = 1, \ldots, n\}; v_i : i = 1, \ldots, m\}$ *be a meta-IFS on* $(\mathcal{X}, d)$, *with mapping* W. *Its unique fixed point,* $A \in (\mathcal{H}(\mathcal{X}), h(d))$, *obeys*

$$A = W(A) = \bigcup_{j=1}^{j=n} w_j(A) \cup \bigcup_{i=1}^{i=m} v_i$$

and is called the attractor of the meta-IFS.

The attractor of an order-m meta-IFS is a metatiling structure. The attractor A is covered by n tiles, which are mappings of itself, and by m external structures. Fortunately, the Collage Theorem applies as before. An example of a metatiling structure is a row of houses vanishing to the horizon; the scene is tiled by two mappings, the first placing a house (an external structure) in the scene and the second taking a house into its neighbour (see Figure 4).

HIERARCHICAL TILING STRUCTURES

Meta-IFSs allow us to describe tiling structures that are hierarchical. A hierarchical metatiling structure (represented by a hierarchical meta-IFS) is a metatiling structure whose external structures are themselves tiling structures. An example of a hierarchical metatiling structure is a row of squares vanishing to the horizon; the scene is tiled by two mappings, the first placing a square (itself tiled by four mappings) in the scene, and the second taking the square into its neighbour (see Figure 5).

IFSs and Tiling Structure Generation

DETERMINISTIC PROCESSES

The attractor (tiling structure) of an IFS is the limit of a sequence of mappings performed on an arbitrary structure [Hutc81].

Process 1 (Attractor) *Let W be the mapping of a simple IFS or of a meta-IFS. The attractor is the unique fixed point $A \in (\mathcal{H}(\mathcal{X}), h(d))$, given by*

$$A = \lim_{n \to \infty} W^{\circ n}(B) \ \ for\ all\ B \in \mathcal{H}(\mathcal{X})$$

The deterministic process generates a converging sequence of intermediate structures (see Figure 6). Starting with an arbitrary structure, the successor structure

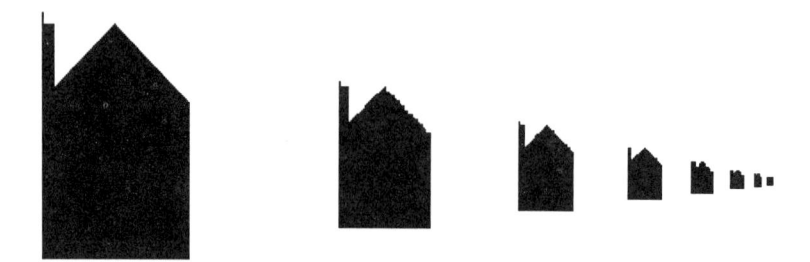

Figure 4. A row of houses.

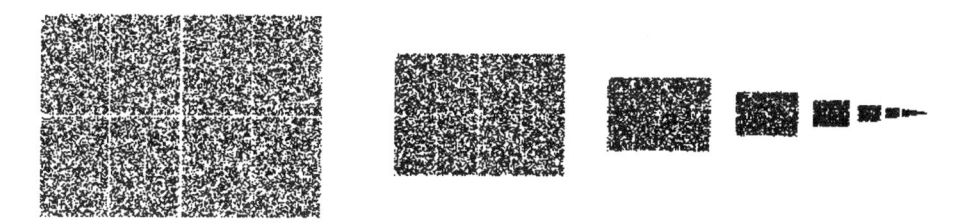

IFSs and the Interactive Design of Tiling Structures   127

Figure 5. A row of squares.

is generated from the union of its predecessor under each of the mappings of the IFS. The external structures upon which the structure of a meta-IFS depends are included without mapping.

RANDOM PROCESSES

Tiling fractal structures (and fractals in general) has a second and equally valid existence as the limit of random processes. Gleick [Glei87] attributes the following analogy to Barnsley:

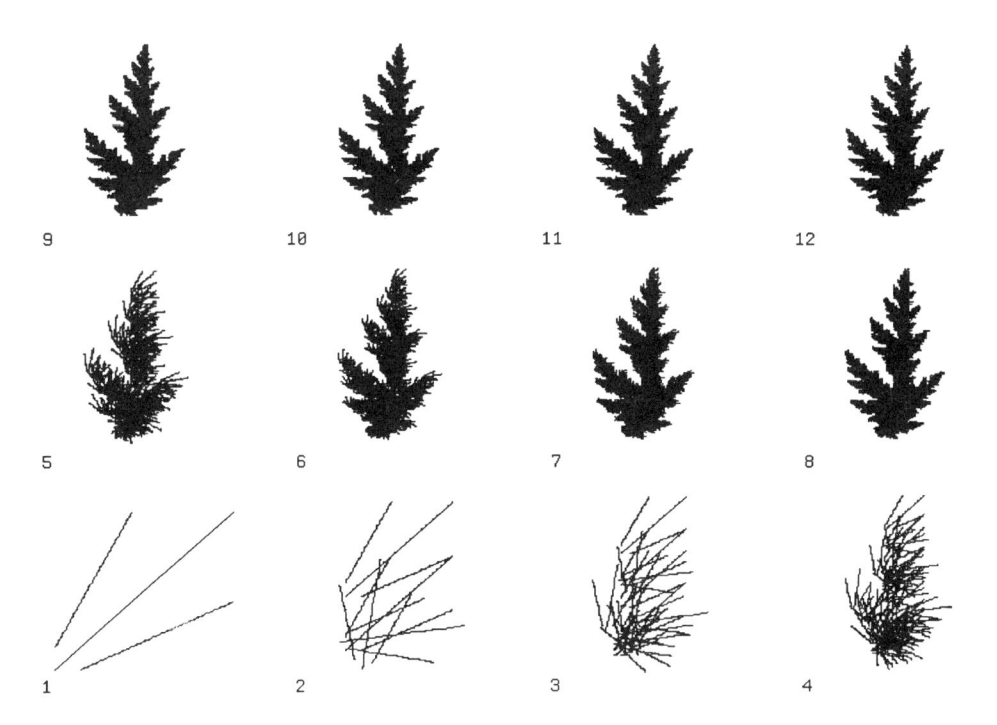

Figure 6. Deterministic process: successive images converging to a leaf.

Imagine a map of Great Britain drawn in chalk on the floor of a room. A surveyor with standard tools would find it complicated to measure the area of these awkward shapes, with Fractal coastlines after all. But suppose one throws grains of rice into the air one by one, allowing them to fall randomly to the floor and counting the grains that land inside the shapes. As time goes on, the result begins to approach the area of the shapes as the limit of a random process.

The limit of a random process can be deterministic and predictable. If this is the case, the limit point does not depend on the randomness; to that extent, chance serves only as a tool. The attractor of a simple IFS can be viewed as the limit of a random process [Barn88a].

Process 2 (Attractor) *Let $\{w_j : j = 1, \ldots, n\}$ be a simple IFS on $(\mathcal{X}, d)$. Choose $x_0 \in (\mathcal{X}, d)$, and at stage $t + 1$ when $x_0, x_1, \ldots, x_t$ have already been determined, randomly choose one of the maps w_j and apply it to x_t to obtain*

$$x_{t+1} = w_j(x_t)$$

The Markov chain $X = \{x_t\}$ is asymptotically stationary and converges in distribution to a unique fixed distribution with support

$$A = W(A) = \bigcup_{j=1}^{j=n} w_j(A)$$

the attractor of the simple IFS.

The random process shown in Figure 7 generates a sequence of points which fall onto (or near) the attractor and once on the attractor stay there. In the initial iterations the process generates a number of garbage points which must be discarded. The rate of convergence of points towards the attractor is determined by the contractivity of the IFS.

One may improve the efficiency of this process by starting the iteration from one of the fixed points of the mappings making up the IFS. Each mapping is a contraction mapping and therefore has a unique fixed point which must lie on the attractor.

Can the attractor of a meta-IFS also be viewed as the limit of a random process? A process for generating the structure of a hierarchical meta-IFS can be constructed from the process above. In the simple process, a 'dancing dot' [Redd88a] moves around the structure according to a sequence of randomly chosen mappings. To generate the structure of a hierarchical meta-IFS, a dancing dot is assigned to the structure of the hierarchical meta-IFS itself, and another dancing dot is assigned to each external structure. Each dot moves independently under the influence of its own IFS. For example, if our hierarchical

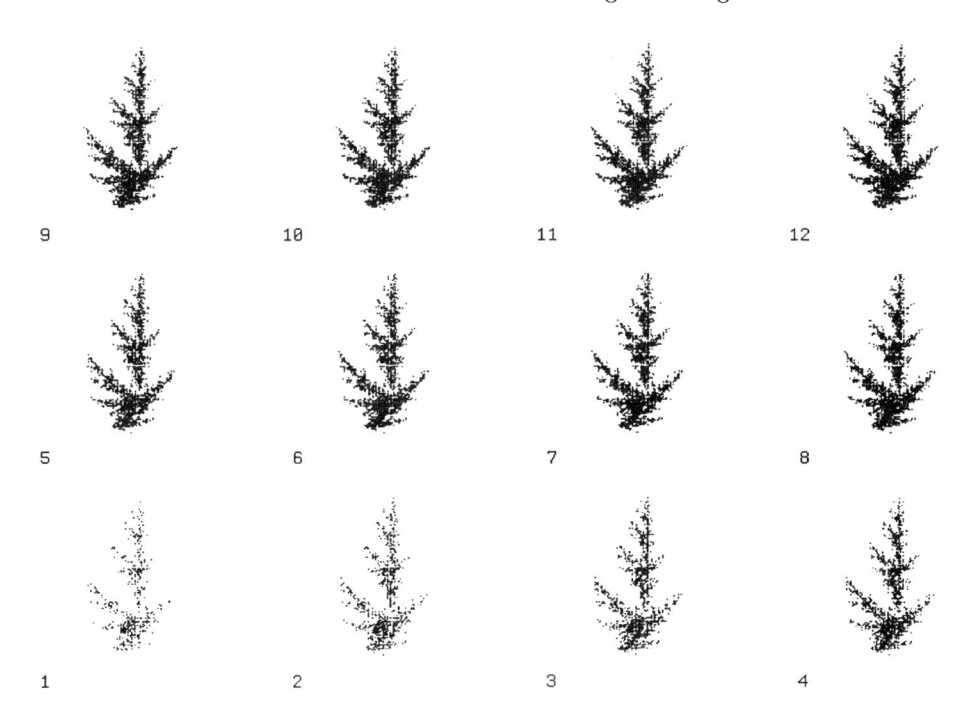

Figure 7. Random process: successive images converging to a leaf.

meta-IFS generates a row of squares vanishing to infinity, then one dot is assigned to the square structure and one to the entire structure. Whenever a mapping is chosen which takes a dot into an external structure (i.e., from the row into the square), the dot takes up the position of the dot in the external structure (i.e., a point in the square). In this way the random processes are *mixed* [Barn88b]. The dancing dot assigned to the structure of the hierarchical meta-IFS visits a sequence of points which fall onto the attractor of the hierarchical meta-IFS.

Process 3 (Attractor) *Let*

$$\{\{w_j : j = 1, \ldots, n\} \quad v_i : i = 1, \ldots, m\}$$

be a hierarchical meta-IFS on $(\mathcal{X}, d)$. *Choose* $x_0 \in (\mathcal{X}, d)$, *and at stage* $t + 1$ *when* $x_0, x_1, \ldots, x_t$ *have already been determined, apply the following random choice*

$$x_{t+1}^0 = \begin{cases} w_j(x_t^0) \\ x_t^i \end{cases}$$

where the Markov chain $X^i = \{x_t^i\}$ *is the outcome of a random process whose limit is* v_i. *The Markov chain* $X^0 = \{x_t^0\}$ *is asymptotically stationary and converges in distribution to a unique fixed distribution with support*

$$A = W(A) = \bigcup_{j=1}^{j=n} w_j(A) \cup \bigcup_{i=1}^{i=m} v_i$$

the attractor of the hierarchical meta-IFS.

The above process can be applied to the general meta-IFS providing random processes can be found for every external structure upon which the meta-IFS depends. Of course, one can always construct a process which generates any given structure and is identical to a stochastic sample of the structure.

A proof of convergence of the random processes is given by Barnsley and Demko [Barn85]. An intuitive explanation requires a consideration of the dynamics of the processes and the 'addressing scheme' defined on a tiling structure by its IFS. We will discuss the mathematics and applications of IFSs viewed as addressing schemes and dynamical systems in a forthcoming paper [Horn90].

IFSs and the DAP

THE PIXEL PLANE

Our description of tiling structure representation and generation using IFSs has so far made no mention of structure storage, manipulation and the rendering of their images using a computer. With the availability of cheap computer memories, by far the most common digital approximation to an image is the *pixel plane* [Newm79]. A pixel plane is a rectangular array of numbers representing logical colours associated with rectangular regions of the image. A pixel plane is an array of *pixels*.

Definition 7 (Pixel plane) *Let $(\mathcal{X}, d)$ be a rectangular region of the Euclidean plane*

$$\mathcal{X} = \{(X, Y) : a \leq X \leq b \quad c \leq Y \leq d\}$$

Partition $\mathcal{X}$ into a grid of $M \times N$ rectangles by dividing the interval $[a, b)$ into M subintervals $[X_{m-1}, X_m)$ for $m = 1, 2, \ldots, M$ where

$$X_m = a + (b - a) m/M$$

and divide the interval $[c, d)$ into N subintervals $[Y_{n-1}, Y_n)$ for $n = 1, 2, \ldots, N$ where

$$Y_n = c + (d - c) n/N$$

Let $P_{m,n}$ denote the rectangle

$$\{(X, Y) : X_{m-1} \leq X < X_m \quad Y_{n-1} \leq Y < Y_n\}$$

and associate a logical colour, $0 \leq C_{m,n} < 2^L$, with each rectangle. Then a pixel plane, P, on $(\mathcal{X}, d)$ is an $M \times N$ array of pixels defined by

$$\{(P_{m,n}, C_{m,n}) : m = 1, \ldots, M \quad n = 1, \ldots, N\}$$

The pixel coordinates index the array, each element of which holds a logical pixel colour.[†]

Definition 8 (Colour map) *Let P be a pixel plane. Then a colour map M is a total function from logical colours to values in some colour space (physical colours).*

We represent a structure's image in a pixel plane firstly by mapping the bounding rectangle of the image onto the region of the Euclidean plane represented by the pixel plane, and secondly by constructing a function mapping pixel coordinates to logical colours. We may, for example, wish to colour all those pixels which contain points in the image of some structure 'black' and all those that do not 'white', in which case our colour-assignment function takes the form

Definition 9 (Colour Assignment) *Let P be a pixel plane of $M \times N$ pixels and 2^L logical colours, representing the same region of the Euclidean plane as occupied by the bounding rectangle of the image, I, of some structure, S. Let C be a total function from pixel coordinates to integers in the range $[0, 2^L)$ defined by*

$$C_{m,n} = \begin{cases} WHITE & \text{if } P_{m,n} \cap I = 0 \\ BLACK & \text{otherwise} \end{cases}$$

for all integers $0 \leq m < M$, $0 \leq n < N$.

ACTIVE MEMORY

The development of VLSI electronics has made possible the combination of processing power and memory on an integrated circuit. The *Distributed Array of Processors* (DAP) [Redd73] is a general purpose massively parallel computing machine which is well suited to the task of digital image generation [Theo89].

The DAP takes the form of an array of interconnected processing elements (PEs), each with local memory attached (see Figure 8). A control unit broadcasts instructions sequentially to the array. Each PE usually has limited capabilities and very limited local autonomy, but taken as a composite whole the array has enormous processing power; algorithms can be designed to exploit this power for almost any problem.

[†]When the pixel plane is displayed on a device a colour map is arbitrarily chosen.

For a two-dimensional processor array, the memory can be visualised as a three-dimensional entity consisting of a stack of two-dimensional array-sized bit-planes. It is the close coupling of processors to memory that has led to the term 'active memory' being applied to such paradigms. Indeed, Active Memory Technology Ltd. design and market the AMT DAPs. Computation on AMT DAPs is characterised by a continuous cycle of activity, with each cycle comprising:

communication of data to the correct processing elements;

arithmetic, logical or comparison operations on data local to each PE.

The purpose of data routing is to pair up within each PE the two operands required for a single operation; the same operation is then performed in every PE but on different data.

The AMT DAP 510 is characterised by:

32×32 array of PEs (1024 in total);

PEs are bit-organised and cycle at 100 nsec;

PEs are 4-connected with wrap-around at the array edges;

row and column 'high-ways';

scalar-matrix global broadcast;

one-bit array-sized mask providing activity control of each PE;

efficient scalar arithmetic and hardware DO-looping provided in MCU;

real-time video board allowing AMT DAP memory visualisation at only two percent overhead.

Parallel Structure Generation and Rendering

Choosing the AMT DAP as our computer and representing the digital image as a pixel plane in AMT DAP memory offers the possibility of real time tiling structure generation and rendering, providing that we identify and exploit the inherent parallelism in the IFS generating processes.

Consider firstly the nature of parallelism in the random generating process for the self-tiling structure of a simple IFS. In the random process a 'dancing dot' moves around the structure according to a sequence of randomly chosen mappings. Two possible methods of exploiting the parallelism in this process are:

dedicating each PE to a dot and bouncing 1024 dots simultaneously under independent random mapping sequences [Redd88a];

dedicating each PE to a dot and bouncing 1024 dots simultaneously under a global random mapping sequence.

The first method simply replicates the random process across all PEs. Each PE should generate its own independent sequence of mappings, but this requires that each has local indexing capabilities. Local indexing is a costly addition to the hardware and not available on the current AMT DAPs. However, the need for local indexing is overstated and may be shown to be unnecessary in the following

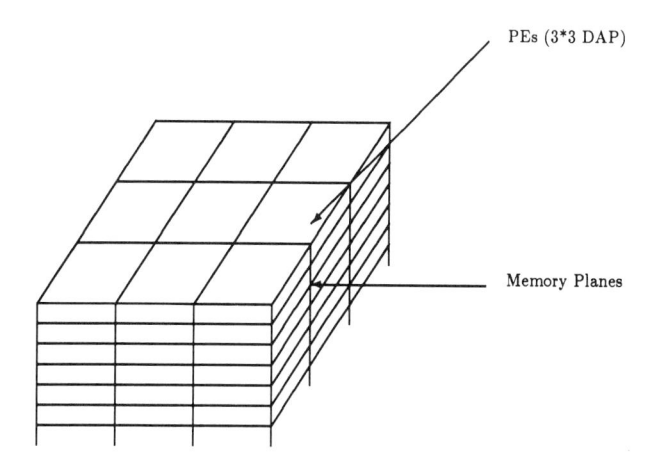

Figure 8. 3D memory/pixel plane storage.

way. Each of the 1024 PEs may generate a semi-independent point sequence in parallel as follows:

Two Monte Carlo selections of the next mapping are made globally.

Each PE uses a local random bit (activity control) to decide which of the two mappings to use.

Each PE performs its chosen mapping.

Points are projected onto the pixel plane (plotted).

A possible subsequence of the globally selected pairs of mappings for a structure with 7 mappings (T1, T2, ..., T7) is:

$$\begin{array}{ccccccc} T1 & T5 & T2 & T2 & T5 & T3 & T5 \ldots \\ T5 & T3 & T5 & T7 & T4 & T2 & T5 \ldots \end{array}$$

and if the corresponding random bits in two PEs are

$$\begin{array}{ccccccc} 1 & 0 & 0 & 1 & 0 & 1 & 1 \ldots \\ 1 & 1 & 0 & 1 & 1 & 0 & 0 \ldots \end{array}$$

then the mapping sequences in the two PEs are

$$\begin{array}{ccccccc} T5 & T5 & T2 & T7 & T5 & T2 & T5 \ldots \\ T5 & T3 & T2 & T7 & T4 & T3 & T5 \ldots \end{array}$$

A small improvement to this method starts all the dots at one of the fixed points of the mappings of the IFS; in this way no points need be discarded. Code implemented on the AMT DAP 510 gives performance approaching one million pixel visits per second.

Alternatively, starting from an arbitrary distribution of dots one can bounce each dot under the same sequence of mappings. This second method generates successive 'clouds' of dots converging towards a single dancing dot (as the mappings of the IFS are all contractive). In order to preserve the parallel nature of the method, one must periodically restart the process with a new arbitrary distribution of dots. This method is faster than the previous method (over one million pixels per second) because of having lower serial overhead. Additionally, if one uses the first method to generate the seed distributions for the second method no points need be discarded.

These methods extend to the meta-IFS and to the generation of tiling structures of arbitrary dimension. However, performance will suffer on a computer which does not have hardware floating point and local indexing capabilities. In particular, the projection of points onto the pixel plane (plotting) is a costly operation.

Secondly, let us consider the nature of parallelism in the deterministic generating process for a plane self-tiling structure (image) described by a simple IFS. In the deterministic generating process an image is repeatedly mapped under a set of mappings. Two observations are central to exploiting the parallelism in the process [Redd88b]:

> a pixel plane maps easily onto the memory of the AMT DAP;
>
> the 2D affine mapping can be regarded as an operation on images rather than an operation on pixel coordinates.

A pixel plane is mapped onto the memory of the AMT DAP by slicing it into AMT DAP-sized 'sheets' and by mapping these sheets into successive matrices in AMT DAP memory. Each PE therefore addresses one pixel from each of the sheets. We regard the 2D affine mapping of an image as a task of movement and interpolation of data mapped onto the 3D memory of the AMT DAP. Performed in this manner, the majority of the work involved in the deterministic process is now performed by the communication network of the array, and the performance of this method should be vastly superior to any other on an AMT DAP.

An efficient implementation of this method on an AMT DAP is helped by two further insights:

> the general affine mapping can be decomposed into simpler mappings acting along the Cartesian axes;
>
> the resolution of the pixel plane representing the image can grow as the process proceeds, starting from an array-sized pixel plane.

The general affine mapping can be decomposed into three simpler mappings: scaling (C), skew (S) and translation (T) (see Figure 9). It can be seen from a construction that the general affine mapping may be rewritten as

$$C_X C_Y S_X R_\theta T_X T_Y$$

where the subscript denotes the usual Cartesian X and Y axes. As shown in Figure 10, the general rotation may also be decomposed into a composition of

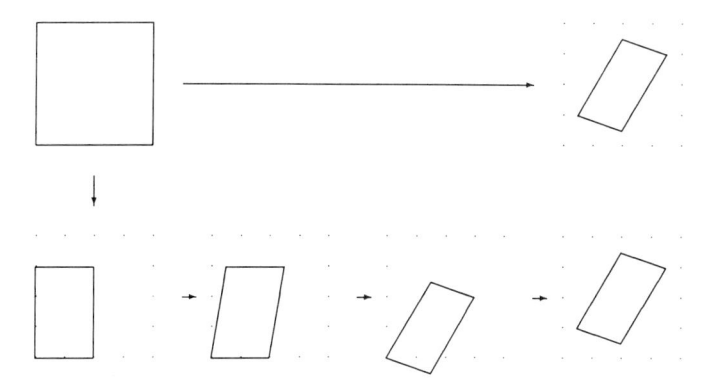

Figure 9. Affine mapping decomposition.

skews [Horn87], e.g.
$$S'_Y S'_X S'_Y$$
Thus, the general affine mapping may be written as
$$C_X C_Y S_X S'_Y S'_X S'_Y T_X T_Y$$
or, equivalently but nonsymmetrically, as
$$C_X C_Y S''_X S''_Y S''_X T_X T_Y$$
Many factors will affect performance of this implementation, e.g.,

> Care must be taken with addressing methods. For example, a skew operation may move data out of the region represented by the pixel plane; this

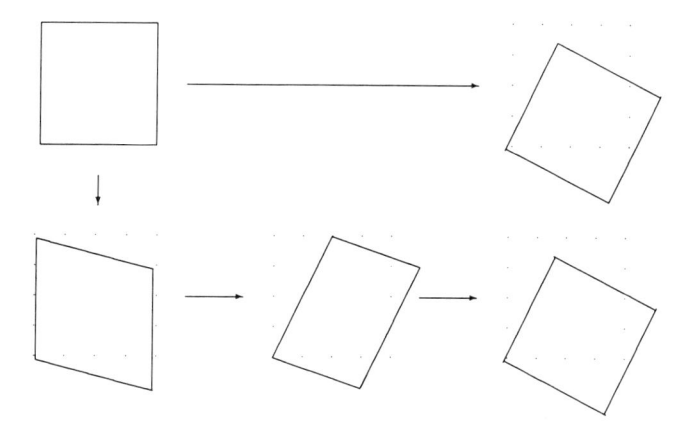

Figure 10. Rotation decomposition.

must either be kept track of by clever addressing or else the image must be bordered by 'white space' into which data can move.

If resolution is increased appropriately through the iterations, then performance is dominated by only the last one or two iterations.

An image with many highly contractive mappings will mean a lot of control work and work on partially filled sheets, but if all mappings are highly contractive few iterations are needed.

For this type of work the programming level and effort is very relevant; a target of 500 cycles per bit-plane of final image is reasonable, but initial high level implementations are more than an order slower. For a 1024×1024 Boolean image, the estimated generation time would be 50 ms.

The deterministic process, when implemented on an AMT DAP, is not only many times faster than the random process but is also more flexible. The random process requires that the external structures of a metatiling structure be the outcome of random processes. The deterministic process applied to the generation of plane metatiling structures can include external images as pixel planes (useful, for example, in mapping real-world 'environment' images as part of a metatiling structure).

The deterministic method may also be applied to 3D 'voxel-based' structures. Researchers at AMT Ltd. are applying the AMT DAP to data visualisation problems in medical physics using some of the techniques outlined above.

IFSs and Tiling Structure Synthesis

We now attempt to answer an important question: Can one synthesise tiling structures by graphical manipulation of a representation of IFSs?

IFSs as Graphical Objects

We confine our discussion to plane self-tiling structures (images) defined in terms of affine mappings contractive in the sense of a special metric. The affine mapping preserves parallel lines and so maps a square into a four-sided figure with opposite sides parallel. The details of our metric are not important, save that a contractive mapping of the unit square in our metric takes the unit square into a parallelogram lying inside (or touching) the unit square.

Given an image defined on the plane, one can draw a bounding box around it and map the box and its contents into the unit square. In this manner, any image can be considered to be contained within the unit square.

A possible graphical representation of our contractive affine mapping is a parallelogram and a unit square.

In Cartesian coordinates, the vertices of the unit square are

$$(0,0), (0,1), (1,1), (1,0)$$

If we label the corresponding vertices of the parallelogram

$$(p_{1x}, p_{1y}), (p_{2x}, p_{2y}), (p_{3x}, p_{3y}), (p_{4x}, p_{4y})$$

the coefficients a, b, c, d, e, f of the affine mapping $x \leftarrow ax+by+e;\ y \leftarrow cx+dy+f$ taking the unit square into the parallelogram are given by

$$f = p_{1y}$$
$$e = p_{1x}$$
$$d = p_{2y} - f$$
$$c = p_{4y} - f$$
$$b = p_{2x} - e$$
$$a = p_{4x} - e$$

If one treats the unit square as a graphical object, the task of specifying an affine mapping involves manipulating a copy of the unit square in order to progressively map it into the desired parallelogram. A suitable interaction device is a mouse or trackball controlling movement of an on-screen cursor. By associating a primitive mapping with each of the vertices of the parallelogram, we provide a flexible means of interaction. A first assignment of mappings to vertices may be

$p_1 \longrightarrow$ translation
$p_2 \longrightarrow$ skew parallel to x-axis
$p_3 \longrightarrow$ rotation about p_1
$p_4 \longrightarrow$ skew parallel to y-axis

Here we utilise the same primitive mappings as were found useful for decomposing the affine mapping for the purposes of image generation in the deterministic generating process. This set of primitives was shown to be sufficient to perform any desired affine mapping; however, it has been found useful to introduce a new primitive, the *metaskew*. One of the four vertices of the parallelogram is designated the 'constraint satisfier', with the other three vertices being free and associated with the metaskew. When a free vertex is moved under the metaskew, the constraint satisfier moves so as to preserve the two parallel line constraints. A single primitive is then sufficient to perform any desired affine mapping, e.g.,

$p_1 \longrightarrow$ metaskew
$p_2 \longrightarrow$ metaskew
$p_3 \longrightarrow$ no mapping—constraint satisfier
$p_4 \longrightarrow$ metaskew

We see that the contractive affine mappings of a simple IFS may be represented as a collection of parallelograms contained within the unit square. How a collection of parallelograms relates to the image they define is made explicit by the following observations which are proved in another context by Barnsley [Barn88a]:

> The IFS image is contained by the parallelograms.
>
> If no parallelograms overlap or touch, the image is disconnected.
>
> If some parallelograms overlap or touch, the image *may* be connected.

These observations suggest that the IFS image provides explicit feedback for the designer (see Figure 11) and so should be generated *in parallel with* the interactive synthesis process.

The methods we have outlined are applicable to all types of IFS, with appropriate modifications to the graphical representations. In the case of a hierarchical meta-IFS, an external image is defined by its own collection of parallelograms. A design tool employing some or all of the features above should provide a flexible environment for the design of tiling structures. Such a tool may find applications in the hands of biological researchers exploring IFSs as a descriptive tool for biological forms, or may be used as the basis for a CAD system.

Case Study: ISIS

ISIS (Interactive System for Image Synthesis) is a prototype interactive plane tiling structure (image) design tool based upon the affine mapping and the random generation process of IFSs. It is implemented on a Sun Microsystems workstation as a prototype for an eventual implementation on an AMT DAP.

ISIS is a window-based system, written in C and running under the SunView window package, constructed to be a design tool for images described by (in the present version) simple IFSs (see Figure 12). It has many of the features we have identified as being necessary for a flexible IFS image design tool:

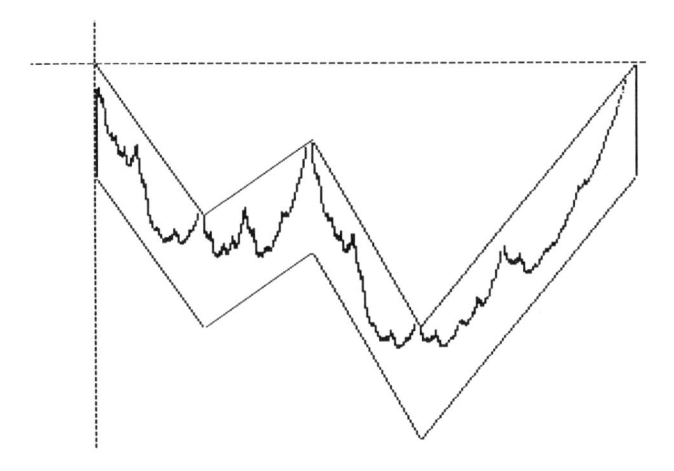

Figure 11. IFS image contained by the parallelograms which define it.

IFSs and the Interactive Design of Tiling Structures 139

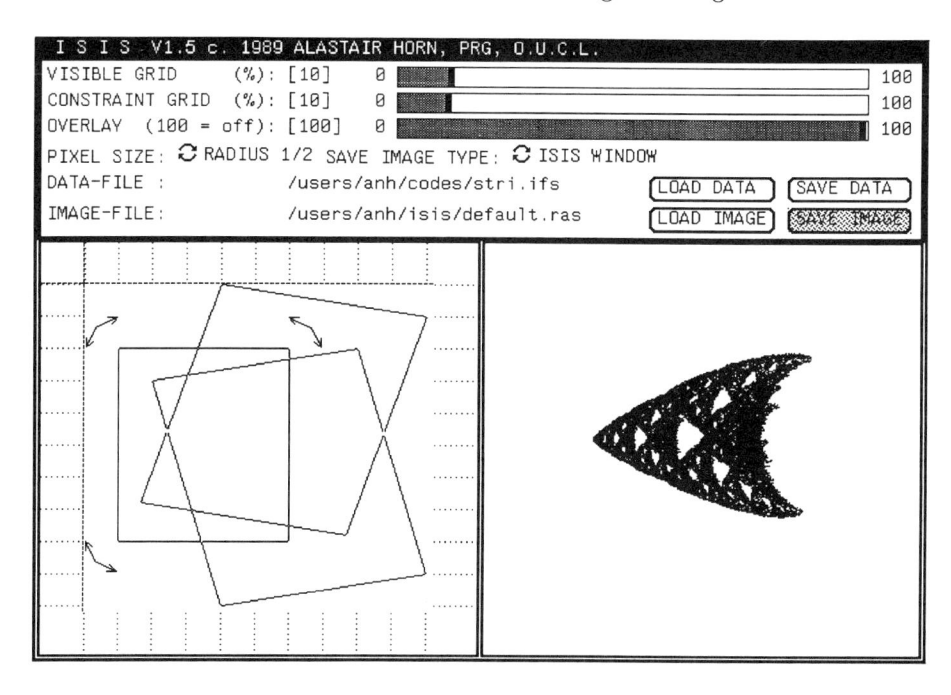

Figure 12. ISIS format.

a 'unit square' drawing surface for IFS design;
a 'unit square' drawing surface for image rendering;
image and design overlay facility;
'parallelism' in the rendering of images and the design process;
mouse interaction with graphical objects;
metaskew vertex to mapping assignment.

We restrict the affine mapping by only allowing affine mappings which take the unit square into itself. This prevents the design of some structures but is not a serious limitation. This restriction allows us to guarantee that the image defined by the IFS will lie in the unit square.

By way of illustration let us consider the design of a leaf such as the one pictured in Figure 13. We start by imagining the image inside the unit square. It may then be seen to be composed of at least four images; its bottom left, bottom right, middle and top are all mappings of the whole. The leaf is a self-tiling structure. The design process is as follows (see Figure 14):

(1) a parallelogram is created/destroyed;
(2) a parallelogram is made current by mouse selection or default;
(3) the mouse is placed near one of the current vertices;
(4) the current parallelogram is metaskewed by dragging a vertex;

140 Alastair N. Horn

Figure 13. Leaf image.

(5) image rendering is restarted;
(6) steps 2 to 5 are repeated until the parallelogram is approximately in its desired position;
(7) steps 1 to 6, or 2 to 6, are repeated until the image tiling has been defined.

An image appears on the image drawing surface as soon as there are two mappings. As each new mapping is added/deleted or an old mapping changed, the

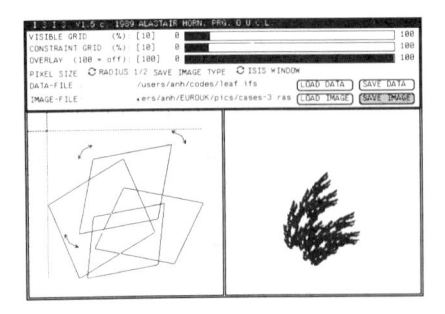

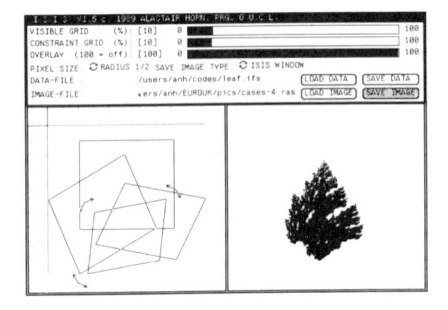

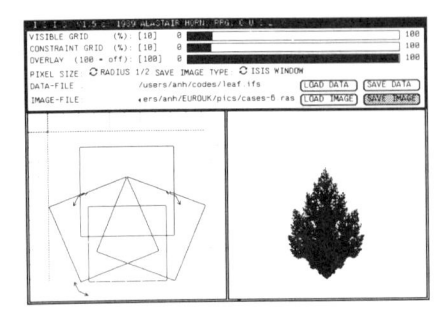

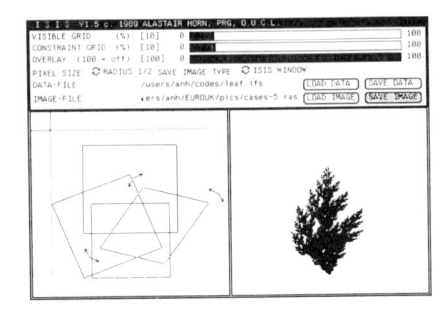

Figure 14. ISIS: Converging to a leaf design.

image rendering is restarted. Rendering proceeds in parallel with the design process at all times, and the generated image may be overlaid on the design drawing surface to aid design at any time, or continuously.

From extensive use of ISIS (see Figure 15) several conclusions may be drawn:

metaskew is an intuitive/flexible primitive;

parallelogram representation of the affine mapping is intuitive;

overlay of the image and design drawing surfaces provides strong feedback.

These conclusions suggest that a parallel implementation should aim to split image rendering and design between a parallel machine and a conventional host, with the host bearing the interaction costs and the parallel machine bearing the rendering costs. This model is that used by the AMT DAP, which is commonly hosted by a Sun or Vax workstation.

The current version of ISIS allows the synthesis of images described by simple IFSs, and provides for the first time an interactive synthesis system for tiling structures and their IFSs. Modifications to allow the design of images described by the full range of IFSs are straightforward.

A new tool developed from ISIS, ISIS+, would utilise nested windows and colour to provide a general IFS picture (multiple image) design system, which would include meta-IFSs and hierarchical meta-IFSs. An area of current research

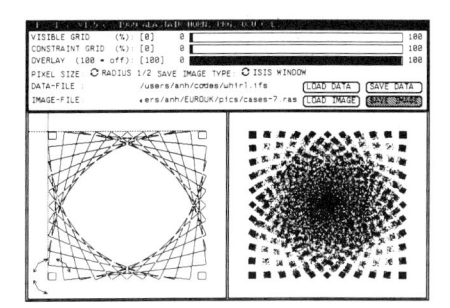

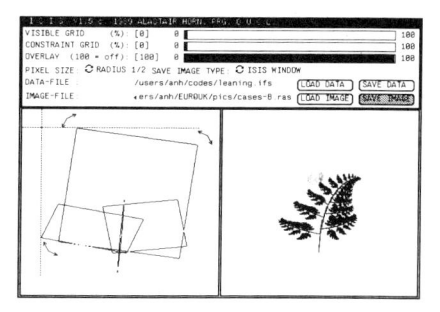

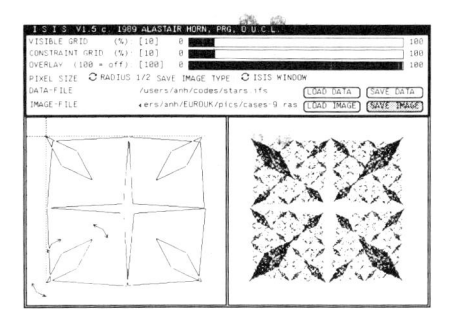

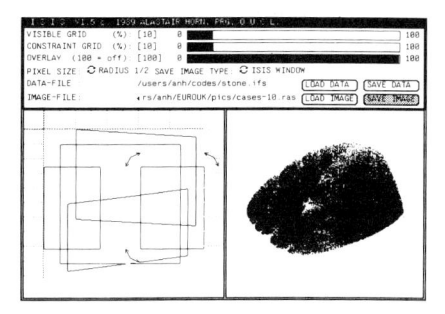

Figure 15. ISIS images: whirl, leaning fern, stars and stone.

is the integration of IFSs and tiling structures with more traditional computer graphics techniques.

Conclusions

In this report we have discussed a representation scheme for tiling structures called *Iterated Function Systems*. We have demonstrated how parallelism may be exploited in the fast generation of tiling structures and their images on the AMT DAP and implemented both random and deterministic algorithms on this massively parallel computer. We have also demonstrated ISIS. ISIS is the first interactive tiling structure design tool based on IFSs.

The foundations of a new area of computer graphics, the computer graphics of tiling structures and IFSs, are now established. We hope to build upon these foundations, extend the work which we have reported upon here and apply IFSs and deterministic fractal geometry to high quality tiling structure design and visualisation for applications in real-time graphics, computer art and CAD.

Acknowledgements. Thanks are due to my academic supervisor, Ian Page, who has proved a constant source of inspiration and a generous giver of sound advice. Thanks also to Mike Godfrey for introducing me to AMT, and to David Hunt, my industrial supervisor, for much support.

Special thanks to Stewart Reddaway and Alan Wilson, consultants with years of DAP experience, who first introduced me to Barnsley's IFSs, who were the first to design and implement IFS algorithms on the AMT DAP, and who have worked with me on the sections of this paper concerned with that massively parallel machine.

And lastly, special thanks to all the people at AMT for producing a range of machines which provide almost unrivaled performance and which above all are a joy to use.

REFERENCES

[Barn85]
　Barnsley, M.F., and Demko, S., Iterated function systems and the global construction of fractals, *Proc. Roy. Soc. London Ser. A*, Vol. 399, pp. 243–275, 1985.

[Barn88a]
　Barnsley, M.F., *Fractals Everywhere*, San Diego: Academic Press, 1988.

[Barn88b]
　Barnsley, M.F., Berger, M.A., and Mete-Soner, H., Mixing Markov chains and their images, *Pre-print*, 1988.

[Falc85]
　Falconer, K.J., *The Geometry of Fractal Sets*, Cambridge tracts in Mathematics 85, 1985.

[Four82]
Fournier, A., Fussell, D., and Carpenter, L., Computer rendering of stochastic models, *CACM*, Vol. 25, pp. 371–384, 1982.

[Glei87]
Gleick, J., *Chaos—Making a New Science*, Cardinal, 1987.

[Grun87]
Grunbaum, B., and Shephard, G.C., *Tilings and Patterns*, New York: W.H. Freeman, 1987.

[Horn87]
Horn A.N., 3-skew image rotation on the DAP, Technical Report ANH001, AMT Ltd., 65 Suttons Park Ave., Reading RG6 1AZ, UK, 1987.

[Horn90]
Horn A.N., Iterated function systems, the parallel progressive synthesis of fractal tiling structures, and their applications to computer graphics, Ph.D. thesis, Oxford University Computing Laboratory, 8–11 Keble Road, Oxford OX1 3QD, UK, 1990.

[Hutc81]
Hutchinson, J.F., Fractals and self-similarity, *Indiana Univ. Jour. of Math.*, Vol. 30, pp. 713–747, 1981.

[Mand68]
Mandelbrot, B.B., and van Ness, J.W., Fractal Brownian motion, fractional noises and applications, *SIAM Rev.*, Vol. 10, pp. 422–437, 1968.

[Mand82]
Mandelbrot, B.B., *The Fractal Geometry of Nature*, New York: W.H. Freeman, 1982.

[Newm79]
Newman, W.M., and Sproull, R.F., *Principles of Interactive Computer Graphics*, New York: McGraw-Hill, 1979.

[Peit86]
Peitgen, H.-O., and Richter, P.H., *The Beauty of Fractals*, Berlin: Springer-Verlag, 1986.

[Peit88]
Peitgen, H.-O., and Saupe, D., *The Science of Fractal Images*, New York: Springer-Verlag, 1988.

[Redd73]
Reddaway, S.F., DAP—a distributed array processor, in *First Annual Symposium on Computer Architecture*, pp. 61–65, Florida, 1973.

[Redd88a]
Reddaway, S.F., and Wilson, A., Regeneration of images from IFS codes on an array processor, in *SIAM annual meeting*, Minneapolis, MN, poster paper, July 1988.

[Redd88b]
Reddaway, S.F., Wilson, A., and Horn, A. N., Fractal graphics and image compression on an SIMD processor, in *Frontiers 88*, Second Symposium on the Frontiers of Parallel Computing, IEEE Computer Society, pp. 265–274, George Mason University, Fairfax, VA, October 1988.

[Suth75]
Sutherland, W.A., *Introduction to Metric and Topological Spaces*, Oxford, UK: Oxford University Press, 1975.

[Theo89]
Theoharis, T., *Algorithms for Parallel Polygon Rendering*, No. 373 in Lecture Notes in Computer Science, Berlin: Springer-Verlag, 1989.

Neural Networks, Learning Automata and Iterated Function Systems

P.C. Bressloff and J. Stark

Abstract

An overview is given of certain underlying relationships between neural networks and Iterated Function Systems. Possible applications to data compression and stochastic learning automata are discussed.

Introduction

Iterated Function Systems (IFSs), which are finite sets of mappings on some metric space, have applications in a number of diverse areas. For example, Barnsley has suggested the use of fractals generated by IFSs as a scheme for data compression in image processing [Barn88a; Barn88b]. The relatively small number of parameters characterising an IFS may be used to specify an extremely complex image. An earlier and very different application of IFSs has been in the analysis of stochastic learning automata [Nare74; Karl53; Norm68; Laks81]. A learning automaton operates in an unknown random environment and updates its action probabilities in accordance with certain reinforcement signals received from the environment. An automaton can thus improve its own performance during operation. The learning rule of the automaton may be described in terms of a random IFS, in which the metric space is the space of action probabilities with Euclidean metric [Norm68]. The performance of the automaton may be related to the invariant measure of the IFS, which often has a fractal structure.

In this paper we discuss how ideas from neural networks may be applied to IFSs and vice versa; this cross-fertilisation may have applications both in data compression and learning automata. We begin by describing a neural network formulation of a class of algorithms for generating images, using IFSs with one neuron per image pixel [Star89]. This opens up the possibility of building dedicated hardware to generate fractal images in fractions of a second. We consider deterministic IFSs in the next section and random IFSs in the section following that one. Just as neural networks may be used to represent IFSs, in the converse direction IFSs provide a useful frame work for studying adaptive neural networks in which learning may be viewed as a random IFS on the space of weights of a network. The limiting behaviour of the network is described by some invariant measure on weight space. Such a measure may have a fractal structure, a point that does not appear to have been realised in previous treatments of neural networks.

In the second part of this paper we illustrate these ideas by formulating a neural model in terms of a random IFS and using this to construct context-dependent learning automata, i.e., such automata learn to associate with each context the particular action which maximises payoff. Our construction is a generalization of the associative search networks of Barto et al. [Bart81; Bart85]. For completeness, we give a brief review of standard learning automata in the fourth section, emphasising the connection with IFSs and fractals, before turning to the neural network features in the last section of the paper.

Deterministic Iterated Function Systems

Let (X, d) be a complete metric space, which we will take to be a bounded subset of R^d, with $d = 2$ for computer graphics applications. An Iterated Function System (IFS) on X is a finite collection of continuous maps $\{f_1, \ldots, f_q\}$. Let $\mathbf{H}(X)$ be the space of compact subsets of X with the Hausdorff metric d_h (see [Fede69; Falc85])

$$d_h(A, B) = \max(d_s(A, B), d_s(B, A)) \qquad A, B \in \mathbf{H}(X)$$

where
$$d_s(A, B) = \max_{x \in A} \min_{y \in B} d(x, y) \qquad A, B \in \mathbf{H}(X)$$

$\mathbf{H}(X)$ is itself a complete metric space. An IFS on X induces a transformation on $\mathbf{H}(X)$ by

$$F(B) = f_1(B) \cup \ldots \cup f_q(B) \qquad \text{for each } B \in \mathbf{H}(X) \tag{1}$$

If the f_α are all contraction mappings on X (a so called hyperbolic IFS), then F is a contraction on $\mathbf{H}(X)$ [Hutc81; Barn88a]. By the contraction mapping theorem, it thus has a unique fixed point A_F (which is a compact subset of X) called the attractor of the IFS. In practical use the f_α are usually taken to be affine, i.e., of the form $f_\alpha(x) = \Lambda_\alpha x + \eta_\alpha$, for $x \in R^2$, where Λ_α is a 2×2 matrix and $\eta_\alpha \in R^2$. Then A_F is encoded by $6q$ real numbers. Many traditional fractals, such as the middle thirds Cantor set and the Sierpinski triangle, may be generated in this way. Define an orbit of an IFS by a sequence $\{x_n\}$ with $x_{n+1} = f_{\alpha_n}(x_n)$, where $\{\alpha_n\}$ is a symbol sequence with $\alpha_n \in \{1, \ldots, q\}$. Then A_F is an attractor in the sense of dynamical systems, since every orbit of an IFS tends to A_F. In particular, for any $x \in A_F$ there is a sequence $\{\alpha_n\}$ such that $x_n \to x$.

Suppose that we now wish to plot the attractor A_F of some hyperbolic IFS on a computer screen. We shall assume that the screen is composed of square pixels and that the horizontal and vertical resolutions of the screen are identical. Let

$X \subset R^2$ be the unit square $\{\mathbf{x} = (x,y) \in R^2 : 0 \leq x, y \leq 1\}$, and assume that we have scaled the IFS appropriately so the $A_F \subset X$. For simplicity we shall represent the space of pixels by the set of points

$$X' = \{\mathbf{x}_{ij} = \left(\frac{i}{\Delta}, \frac{j}{\Delta}\right) \in X : i, j \in \{0, 1, \ldots, \Delta - 1\}\} \subset X \qquad (2)$$

where Δ is the resolution of the screen. It may be shown that taking the $(i,j)^{\text{th}}$ pixel to be the square $[i/\Delta, (i+1)/\Delta] \times [j/\Delta, (j+1)/\Delta]$ rather than the point $\mathbf{x}_{ij}$ is only marginally more accurate but much more demanding to implement (see [Star89]).

One method for drawing an approximation of A_F is based on the iteration of one of more discrete orbits. That is, choose an initial point $\mathbf{x}_0$ and plot the set

$$A(\mathbf{x}_0, N) = \bigcup_{\alpha \in \Omega_q^N} f_{\alpha_N} \ldots f_{\alpha_1}(\mathbf{x}_0)$$

where Ω_q^N is the set $\{\boldsymbol{\alpha} = (\alpha_1, \ldots, \alpha_N) : \alpha_n \in \{1, \ldots, q\}\}$ of symbol sequences of length N on q symbols. If we choose N sufficiently large, the set $A(\mathbf{x}_0, N)$ will be a good approximation of A_F. A more efficient procedure is to take $\mathbf{x}_0$ to be a fixed point of one of the f_α, so that $\mathbf{x}_0 \in A_F$. Then we can also plot all the intermediate points $f_{\alpha_n} \ldots f_{\alpha_1}(\mathbf{x}_0)$ on the orbit. Furthermore, more than one initial point $\mathbf{x}_0$ may be used for iteration. This is precisely one of the methods used by Prusinkiewicz and Sandness [Prus88] to draw Koch curves. The orbit iteration method may be extended to random IFSs, as described in the next section, and is by far the most common algorithm for generating images from an IFS. However, such an approach is not very suitable for parallel implementation. For although one may iterate more than one orbit simultaneously, iterating N orbits in one iteration is no substitute for iterating one orbit for N iterations. Thus, in practice the number of processors that orbit iteration can usefully employ is much smaller than the number of screen pixels.

An alternative method for generating IFS images is based on the direct calculation of an operator in image space, which, for a deterministic IFS, is the mapping F of Eq. (1). The basic idea is to choose an initial $A_0 \in \mathbf{H}(X)$ and define

$$A_{n+1} = f_1(A_n) \cup \ldots \cup f_q(A_n) \qquad (3)$$

Then the sequence A_n will tend to A_F as $n \to \infty$, irrespective of the choice of A_0. We thus draw A_N for some sufficiently large N, determined by the contractivity factors of the f_α. An example of such a deterministic algorithm is given by Barnsley [Barn88a]. This direct approach is suitable for parallel implementations either on an array processor as suggested by Reddaway [Redd88], or on a dedicated neural network as proposed by Stark [Star89]. In either case, the maximum number of useful processors or neurons is equal to the number of pixels on the computer screen, and thus this is potentially a very fast algorithm. We shall now describe the neural network formulation of a deterministic IFS based upon this approach, which will be extended to a random IFS in the next section.

The first step is to obtain an approximation of f_α by a map on the pixel space X' defined in Eq. (2). For $u \in R$, let $\lceil u \rceil$ denote the smallest integer such that $\lceil u \rceil \geq u$. Then taking $\lceil u - 1/2 \rceil$ corresponds to the usual action of rounding a real number to the nearest integer. Introduce the mapping $\Theta : X \to X'$ such that

$$\Theta(x, y) = \left(\frac{\lceil x\Delta - 1/2 \rceil}{\Delta}, \frac{\lceil y\Delta - 1/2 \rceil}{\Delta} \right) \tag{4}$$

Thus $\Theta(\mathbf{x})$ for $\mathbf{x} \in X$ is the nearest point in X' to $\mathbf{x}$. Hence, given $\mathbf{x}_{ij} \in X'$, the best approximation to $f_\alpha(\mathbf{x}_{ij})$ in X' is $\Theta(f_\alpha(\mathbf{x}_{ij}))$. This gives an approximation of F restricted to $B \in \mathbf{H}(X')$ by

$$F'(B) = \Theta \circ f_1(B) \cup \ldots \cup \Theta \circ f_q(B) \tag{5}$$

Then, as in Eq. (3), we take an arbitrary $A_0 \in \mathbf{H}(X')$, define the sequence A_n by $A_{n+1} = F'(A_n)$ and draw A_n for some sufficiently large N. The iteration F' can be computed in a finite manner, since each A_n consists of a finite collection of points. Moreover, it has a simple realisation as a neural network. To show this, note that each A_n can be uniquely specified by the set of binary valued variables $\{a_{ij} : i, j \in \{1, \ldots, \Delta - 1\}\}$ where

$$a_{ij}(n) = \begin{cases} 1 & \text{if } \mathbf{x}_{ij} \in A_n \\ 0 & \text{if } \mathbf{x}_{ij} \notin A_n \end{cases}$$

If we define

$$\omega_{iji'j'} = \begin{cases} 1 & \text{if } \Theta \circ f_\alpha(\mathbf{x}_{i'j'}) = \mathbf{x}_{ij} \quad \text{for some } \alpha \\ 0 & \text{otherwise} \end{cases} \tag{6}$$

then the dynamics of $a_{ij}(n)$ is given by

$$a_{ij}(n+1) = \theta \left(\sum_{i'j'} \omega_{iji'j'} \, a_{i'j'}(n) \right) \tag{7}$$

where θ is the step function

$$\theta(x) = \begin{cases} 1 & \text{if } x > 0 \\ 0 & \text{if } x \leq 0 \end{cases} \tag{8}$$

If we interpret a_{ij} as the state of a binary neuron, then Eq. (7) is precisely the dynamics of a binary threshold network, with Δ^2 neurons indexed by (i, j) and connection weights given by the $\omega_{iji'j'}$. Note that each neuron's threshold has been set to zero, but we could equally well have used any value V such that $0 \leq V < 1$. The above network is particularly simple, as the weights only have values 0 and 1, and each neuron is just an OR gate, i.e., it fires if any of its inputs are on. We remark:

The output of each neuron is connected to the input of, at most, q other neurons. Therefore, although there is an extremely large number of weights $\omega_{iji'j'}$ (e.g., about 10^{12} for a high resolution screen), the vast majority will be zero. The resulting network is sparsely connected, making hardware implementation feasible. Furthermore, sparse matrix methods may be used to store the nonzero weights in simulation.

The formulation is easily extended to include a condensation set, D, by setting $a_{ij}(n) = 1$ for all n and (i, j) such that $\mathbf{x}_{ij} \in D$.

Since we are working with an approximation of the real IFS, this approach is potentially less accurate than orbit iteration methods. In particular, F' will in general no longer be a contraction of $\mathbf{H}(X')$, and hence we can no longer guarantee convergence to a unique fixed point of the sequence A_n. However, it can be shown that the limiting behaviour is confined to a bounded neighbourhood of the real fixed point A_F of F. More precisely (see [Star89]), define a neighbourhood of A_F by

$$\mathbf{V} = \left\{ B \in \mathbf{H}(X') : d_h(A_F, B) \leq \frac{1}{\sqrt{2}\Delta(1-\lambda)} \right\}$$

where $\lambda = \max_\alpha \lambda_\alpha$, and λ_α is the contraction ratio of f_α

$$\lambda_\alpha = \sup_{\mathbf{x},\mathbf{y} \in X} \frac{d(f_\alpha(\mathbf{x}), f_\alpha(\mathbf{y}))}{d(\mathbf{x},\mathbf{y})}$$

Then for every $A_0 \in \mathbf{H}(X')$, $\mathbf{V}$ contains the ω-limit set under iteration by F', i.e., if $A_n = F'^n(A_0)$, then A_n will tend to $\mathbf{V}$. Note that the size of $\mathbf{V}$ depends on the pixel size $1/\Delta$ and the contraction λ. If $\lambda < (1 - 1/\sqrt{2})$, then the error in the approximate image will be at most one pixel, i.e., $d_h(A_F, A_n) \leq 1/\Delta$. Also note that orbit iteration methods require extremely long orbits to achieve their full potential accuracy.

Random Iterated Function Systems

A deterministic IFS can only generate a black and white image. To define an IFS which generates colour or grey scale images, we assign a probability p_α to each map f_α, with $\sum_\alpha p_\alpha = 1$ and $p_\alpha > 0$. For simplicity we take the p_α to be constants. (On the other hand, in the next section we shall consider stochastic learning automata, whose associated IFSs have state-dependent distributions $p_\alpha(x)$). Let $\mathbf{B}(X)$ be the Borel σ-field of X (see, e.g., [Fede69] or [Falc85]) and $\mathbf{P}(X)$ the space of probability measures on $\mathbf{B}(X)$. Define a Markov transition probability or kernel by

$$K(x, B) = \sum_\alpha p_\alpha \mathcal{X}_B(f_\alpha(x)) \tag{9}$$

which is the probability of transfer from $x \in X$ into the Borel set B. Here $\mathcal{X}_B$ is the indicator function

$$\mathcal{X}_B(x) = \begin{cases} 1 & \text{if } x \in B \\ 0 & \text{otherwise} \end{cases}$$

The kernel K induces a Markov operator M on $\mathbf{P}(X)$ by

$$M(\mu)(B) = \int_X K(x,B) d\mu(x) = \sum_\alpha p_\alpha \mu(f_\alpha^{-1}(B)) \qquad \mu \in \mathbf{P}(X) \qquad (10)$$

and the last equality follows from the fact that the associated IFS has constant probabilities p_α.

If the space X is compact, we can endow $\mathbf{P}(X)$ with the Hutchinson metric

$$d_H(\mu,v) = \sup_{\phi \in C} \left\{ \int_X \phi d\mu - \int_X \phi dv \right\} \qquad \mu,v \in \mathbf{P}(X)$$

where

$$C = \{ \phi : X \to R : \mid \phi(x) - \phi(y) \mid \leq d(x,y) \qquad \text{for all } x, y \in X \}$$

Then $(\mathbf{P}(X), d_H)$ is a complete metric space [Hutc81]. Moreover, it can be shown that if the f_α are contractions on X, then M is a contraction on $(\mathbf{P}(X), d_H)$ [Hutc81; Barn88]. The fixed point μ_F is called the invariant measure of the IFS. Such measures are particularly useful for computer graphics applications. Clearly they yield grey scale images directly, and, by superimposing three such measures in the three primary colours, they can be used to give colour images. We remark that the support of μ_F is the attractor A_F of the corresponding deterministic IFS. Define orbits of an IFS as in the previous section on deterministic IFSs. Then, for almost every symbol sequence $\{\alpha_n\}$, the visiting frequency of the orbit $\{x_n\}$ to a measurable set B is $\mu_F(B)$ (see [Elto87])

$$\lim_{N \to \infty} \frac{\#\{x_n \in B : 1 \leq n \leq N\}}{N} = \mu_F(B) \qquad B \subset \mathbf{B}(X)$$

We now discuss how to plot the invariant measure μ_F of some hyperbolic random IFS on a computer screen along similar lines to the previous section. We shall take the pixel representation as defined in Eq. (2). Again, we can divide drawing algorithms into those which depend on the iteration of one or more discrete orbits, and those which work directly with an operator on image space, which for a random IFS is the Markov operator M. Orbit iteration methods involve choosing an f_α randomly at each iteration. Thus, we generate the orbit $\mathbf{x}_{n+1} = f_{\alpha_n}(\mathbf{x}_n)$ where f_{α_n} is chosen with probability p_{α_n}. For N sufficiently large

$$\mu_N(B) - \frac{\#\{\mathbf{x}_n \in B : 1 \leq n \leq N\}}{N} \qquad (11)$$

will be a good approximation to $\mu_F(B)$. The alternative approach is to choose an initial $\mu_0 \in \mathbf{P}(X)$ and define a sequence of measures by

$$\mu_{n+1} = M(\mu_n) = \int_X K(\mathbf{x},.) d\mu_n(\mathbf{x}) \tag{12}$$

As in the deterministic case, this method is more suitable for parallel implementation than orbit iteration. In particular, a neural network formulation of a random IFS may be constructed as detailed below (also see [Star89]).

We proceed by defining an operator M' on $\mathbf{P}(X')$, where X' is the pixel space (see Eq. 2). We can write any measure $\mu \in \mathbf{P}(X')$ as

$$\mu = \sum_{\{i,j\}} \mu_{ij} \delta_{\mathbf{x}_{ij}}$$

where $\delta_{\mathbf{x}_{ij}}$ is the δ measure at $\mathbf{x}_{ij}$. Thus, it is sufficient to specify M' by the values $[M'(\mu)]_{ij}$. Introduce the kernel

$$K'(\mathbf{x}_{i'j'}, \mathbf{x}_{ij}) = \sum_\alpha p_\alpha \delta_{\mathbf{x}_{ij}}(\Theta \circ f_\alpha(\mathbf{x}_{i'j'})) \tag{13}$$

where Θ is given in Eq. (4). Note that $p_\alpha \delta_{\mathbf{x}_{ij}}(\Theta \circ f_\alpha(x_{i'j'}))$ is the probability that the nearest point in X' to $f_\alpha(\mathbf{x}_{i'j'})$ is $\mathbf{x}_{ij}$. Then K' induces the Markov operator M' acting on $\mathbf{P}(X')$

$$[M'(\mu)]_{ij} = \sum_{i'j'} K'(\mathbf{x}_{i'j'}, \mathbf{x}_{ij}) \mu_{i'j'} \tag{14}$$

If we set

$$\omega_{iji'j'} = K'(\mathbf{x}_{i'j'}, \mathbf{x}_{ij}) \tag{15}$$

Eq. (14) may be written in the dynamical form

$$\mu_{ij}(n+1) = \sum_{i'j'} \omega_{iji'j'} \mu_{i'j'}(n) \tag{16}$$

which again has a simple interpretation as a neural network, this time as an analogue linear net. Each pixel corresponds to a single neuron whose state at time n is given by $\mu_{ij}(n) \in [0,1]$. The value of $\mu_{ij}(n)$ determines the intensity of the $(i,j)^{\text{th}}$ pixel on the computer screen. Note the following:

One advantage of this formalism is the way a random, possibly nonlinear, algorithm has been replaced by a deterministic linear one (see Eq. (16)).

As with the deterministic case, it is necessary to take into account the fact that we are working with an approximation to the real IFS. For a random IFS, this means that M' is not necessarily a contraction on $\mathbf{P}(X')$, and μ_n

will not in general converge to a unique fixed point. Nevertheless, it may be shown [Star89] that the ω-limit set of any orbit under M' is

$$\mathbf{W} = \left\{\mu \in \mathbf{P}(X') : d_H(\mu_F, \mu) \leq \frac{1}{\sqrt{2}\Delta(1-\lambda)}\right\} \quad (17)$$

Moreover, since $\mathbf{P}(X')$ is finite-dimensional and M' is a linear mapping of $\mathbf{P}(X')$ to itself, we may apply the Perron-Frobenius theorem [Sene81] to analyse the fixed and periodic points of the ω-limit set of M'.

Condensation sets may be added by taking an affine rather than linear net.

Stochastic Learning Automata

In the previous two sections we considered the application of IFSs in computer graphics. We now turn to an earlier use of random IFSs arising in learning automata. For a general review of learning automata, see the work of Narendra and Thathachar [Nare74] and references therein. In this section, we briefly discuss standard automata theory and make explicit the connection with IFSs as developed by Karlin [Karl53] and Norman [Norm68; Norm72]. In the next section we will describe an extension of the theory using ideas from neural networks.

A learning automaton operates in an unknown random environment with which it is connected in a feedback loop. The environment receives each action emitted by the automaton and produces a response, which acts as input to the automaton, that evaluates the suitability of the action. The evaluative feedback is usually a binary signal, indicating either success (reward) or failure (punishment) and is probabilistic in nature. Learning occurs if the automaton tends to increase its likelihood of success, the optimal state being that the automaton emits only the action corresponding to the largest probability of success. More formally

(a) At each time-step n, the automaton selects an action y_n from the finite set $Y = \{y^{(1)}, \ldots, y^{(r)}\}$. The action is chosen randomly with

$$\mathbf{Pr}\{y_n = y^{(i)}\} = p_n^{(i)} \quad (18)$$

The state of the automaton is the probability vector

$$\mathbf{s}_n = (p_n^{(1)}, \ldots, p_n^{(r)})^T \qquad \sum_{i=1}^{r} p_n^{(i)} = 1$$

The corresponding state space S is the unit $(r-1)$-dimensional simplex. We shall assume that S is a metric space with respect to some metric d. Then (S, d) is complete. The appropriate choice for d will depend on which learning rule is adopted (see below). For example, in the case of linear models it is sufficient to take the Euclidean metric.

Neural Networks, Learning Automata and Iterated Function Systems   153

(b) The environment receives y_n as input and responds by sending an evaluative signal $b_n \in \{0,1\}$, where 1 and 0, respectively, correspond to success and failure. The environment is characterised by a set of success probabilities $\{\pi^{(1)}, \ldots, \pi^{(r)}\}$, where

$$\pi^{(i)} = \mathbf{Pr}\{b_n = 1 \mid y_n = y^{(i)}\} \tag{19}$$

The probabilities $\pi^{(i)}$ are not known *a priori*. For simplicity, we shall take the $\pi^{(i)}$ to be independent of time; the environment is said to be stationary.

(c) Define an event e_n as the pair

$$e_n = (i_n, b_n) \in E \tag{20}$$

where
$$i_n = \begin{cases} 1 & \text{if } y_n = y^{(1)} \\ \ldots & \ldots \\ r & \text{if } y_n = y^{(r)} \end{cases}$$

and E is the set of possible events $\{(i,b) : 1 \leq i \leq r,\ b \in \{1,0\}\}$. Each event represents the pairing of an action with its associated environmental response. Given a state $\mathbf{s} = (p^{(1)}, \ldots, p^{(r)}) \in S$, from (a) and (b) the probability of an event $e = (i, b)$ is

$$\phi_e(\mathbf{s}) = p^{(i)}[\pi^{(i)}b + (1 - \pi^{(i)})(1 - b)] \tag{21}$$

The ϕ_e are probability distributions on S, with $\phi_e : S \to [0,1]$ such that $\sum_{e \in E} \phi_e(\mathbf{s}) = 1$.

(d) A learning algorithm for the automaton is a set of continuous mappings on S, $\{f_e : S \to S : e \in E\}$. When the maps are linear in the probabilities $p^{(i)}$, the automaton is said to be linear. If the state of an automaton at a particular time is $\mathbf{s}$, the occurrence of an event e, which from (c) has a probability $\phi_e(\mathbf{s})$, effects a change of state $\mathbf{s} \to f_e(\mathbf{s})$. This leads to a dynamical equation for updating action probabilities

$$\mathbf{s}_{n+1} = f_{e_n}(\mathbf{s}_n) \tag{22}$$

where f_{e_n} is chosen at random with probability $\phi_{e_n}(\mathbf{s}_n)$.

From the above description we see that the basic operation carried out by a learning automaton is the updating of the action probabilities on the basis of the responses from the environment. It remains to determine whether the updating leads to improved performance of the automaton. One quantity useful in judging the behaviour of an automaton is the probability of success at step n, which is

$$Q_n = \sum_{i=1}^{r} \pi^{(i)} p_n^{(i)} \tag{23}$$

If no *a priori* information is available and the actions are chosen with equal probability, the average success is

$$Q^0 = \sum_{i=1}^{r} \frac{\pi^{(i)}}{r}$$

Roughly speaking, an improvement in performance occurs if asymptotically Q_n tends to be greater than Q^0. More precisely [Nare74], a learning automaton is said to be

expedient if $\lim_{n\to\infty} \mathbf{E}(Q_n) > Q^0$;

optimal if $\lim_{n\to\infty} \mathbf{E}(Q_n) = \pi^l$ where $\pi^l = \max_{i} \{\pi^{(i)}\}$;

ϵ-optimal if $\lim_{n\to\infty} \mathbf{E}(Q_n) > \pi^l - \epsilon$ can be obtained for any $\epsilon > 0$ by a suitable choice of learning algorithm parameters.

Here $\mathbf{E}$ denotes expectation with respect to the random action probabilities $p_n^{(i)}$.

The limiting behaviour of a learning automaton may be described in terns of the invariant measure of a random IFS. For it is clear that the set $\{(f_e, \phi_e) : e \in E\}$ defines a random IFS in the metric space (S, d). Following the earlier section on random IFSs, we introduce the stochastic kernel

$$K(\mathbf{s}, B) = \sum_{e \in E} \phi_e(\mathbf{s}) \mathcal{X}_B(f_e(\mathbf{s})) \qquad B \in \mathbf{B}(S) \tag{24}$$

Analogous to Eq. (12), by choosing an initial $\mu_0 \in \mathbf{P}(S)$ we have a sequence of measures induced by K

$$\mu_{n+1} = \int_S K(s, .) d\mu_n(\mathbf{s}) \tag{25}$$

The asymptotic behaviour of the automaton may then be formulated in terms of the limit of the sequence $\{\mu_n\}$, assuming it converges weakly in distribution, since

$$\lim_{n\to\infty} \mathbf{E}\{Q_n\} = \lim_{n\to\infty} \int_S \left[\sum_i \pi^{(i)} p^{(i)}\right] d\mu_n(\mathbf{s}) \tag{26}$$

Introduce the n-step transition probability for $n \geq 1$ by

$$K^{(n)}(\mathbf{s}, B) = \sum_{e_1,\ldots,e_n} \phi_{e_1}(\mathbf{s})\phi_{e_2}(f_{e_1}(\mathbf{s}))\ldots\phi_{e_n}(f_{e_1\ldots e_{n-1}}(\mathbf{s})) \mathcal{X}_B(f_{e_1\ldots e_n}(\mathbf{s})) \tag{27}$$

where $f_{e_1\ldots e_n} = f_{e_n} f_{e_{n-1}} \ldots f_{e_1}$. Also let $K^{(0)}(\mathbf{s}, B) = \mathcal{X}_B(\mathbf{s})$. Then

$$\mu_n = \int_S K^{(n)}(\mathbf{s}, .) d\mu_0(\mathbf{s})$$

and we may discuss the asymptotic behaviour of the automaton in terms of the limit of the sequence $\{K^{(n)}\}$. A sequence $\{K^{(n)}\}$ of kernels is said to converge to a kernel K_∞ if for any Borel subset B of S and any $\epsilon > 0$ there is an integer N such that
$$K_\infty(\mathbf{s}, B^0) - \epsilon \leq K^{(n)}(\mathbf{s}, B) \leq K_\infty(\mathbf{s}, \bar{B}) + \epsilon$$
for all $n \geq N$ and $\mathbf{s} \in S$. Here, B^0 is the interior and $\bar{B}$ the closure of B.

To proceed further, it is necessary to take into account the fact that the ϕ_e are state-dependent distributions rather than constants, as assumed for the analogous p_α in the previous section. Random IFSs with variable probabilities have been discussed by Karlin [Karl53], Norman [Norm68] (in the context of mathematical learning theory), and more recently by Barnsley et al. [Barn85] and Elton [Elto87]. In particular, Norman [Norm68] distinguishes two important types of limit for the sequence $\{K^{(n)}\}$. To describe these limits it is useful to introduce some extra notation. We shall assume that $\{(f_e, \phi_e) : e \in E\}$ is an arbitrary random IFS on some metric space (S, d). We shall return to the specific example of learning automata below. Let $T_n(\mathbf{s})$ denote the set of values that $\mathbf{s}_n$ has with positive probability, given $\mathbf{s}_0 = \mathbf{s}$. Thus

$$T_n(\mathbf{s}) = \{\mathbf{s}' : K^{(n)}(\mathbf{s}, \{\mathbf{s}'\}) > 0\} \qquad (28)$$

Define an absorbing state as one which satisfies $T_1(\mathbf{s}) = \{\mathbf{s}\}$, i.e., once entered it cannot be left. Next, set $\bar{d}(A, B)$ to be the minimum distance between any two subsets A and B of S
$$\bar{d}(A, B) = \min_{\substack{\mathbf{s} \in A \\ \mathbf{s}' \in B}} d(\mathbf{s}, \mathbf{s}')$$

Finally, define a (strictly) distance diminishing model to be a random hyperbolic IFS $\{(f_e, \phi_e) : e \in E\}$, i.e., the f_e are contracting, which acts on a compact metric space (S, d) and whose dynamics is a Markov process generated by the stochastic kernel (Eq. 24). Norman [Norm68] has shown that

(a) If a distance diminishing model satisfies
$$\lim_{n \to \infty} \bar{d}(T_n(\mathbf{s}), T_n(\mathbf{s}')) = 0 \qquad \text{for all } \mathbf{s}, \mathbf{s}' \in S \qquad (29)$$
then the asymptotic distribution $K_\infty(\mathbf{s}, .) = K_\infty(\mathbf{s})$ is independent of the initial state $\mathbf{s}$, and $K^{(n)}$ converges uniformly to K_∞. The associated invariant measure μ_F is ergodic. Note that for Eq. (29) to hold there can be at most one absorbing state.

(b) Suppose that a distance diminishing model has $N > 1$ absorbing states $\mathbf{a}_1, \ldots, \mathbf{a}_N$ such that, for any $\mathbf{s} \in S$, there is some $\mathbf{a}_{j(\mathbf{s})}$ for which
$$\lim_{n \to \infty} \bar{d}(T_n(\mathbf{s}), \{\mathbf{a}_{j(\mathbf{s})}\}) = 0 \qquad (30)$$
Then the sequence of states $\{\mathbf{s}_n\}$ converges with probability one to a random absorbing state $\mathbf{s}_\infty$. Given an initial state $\mathbf{s}$, the kernel $K^{(n)}$

converges uniformly to K_∞ such that $K_\infty(\mathbf{s},.)$ assigns a probability $\gamma_i(\mathbf{s})$ to $\mathbf{a}_i$, i.e.
$$\gamma_i(\mathbf{s}) = \mathbf{Pr}\{\mathbf{s}_\infty = \mathbf{a}_i \,|\, \mathbf{s}_0 = \mathbf{s}\}$$
Since the asymptotic behaviour depends on the initial conditions and is thus not unique, the model is said to be nonergodic, or absorbing.

We now interpret these results in terms of the limiting behaviour of a learning automaton. The mode of convergence (a) occurs typically in the case of expedient learning schemes, such as the linear reward–penalty scheme (L_{R-P}): for $e = (i, 1)$
$$p_{n+1}^{(i)} = p_n^{(i)} + \alpha(1 - p_n^{(i)})$$
$$p_{n+1}^{(j)} = (1 - \alpha) p_n^{(j)} \qquad j \neq i \tag{31}$$

and for $e = (i, 0)$
$$p_{n+1}^{(i)} = (1 - \beta) p_n^{(i)}$$
$$p_{n+1}^{(j)} = \frac{\beta}{r-1} + (1 - \beta) p_n^{(j)} \qquad j \neq i \tag{32}$$

where $0 < \alpha, \beta < 1$. The basic idea behind the scheme is simple. If the learning automaton selects an action $y_n = y^{(i)}$ and a reward occurs, then the action probability $p_n^{(i)}$ is increased and all the other components $p_n^{(j)}$, $j \neq i$ are decreased. For a punishment response, $p_n^{(i)}$ is decreased and the other components increased. The second mode of convergence, (b), typically corresponds to ϵ-optimal schemes. An example of this is the linear reward–inactivity scheme (L_{R-I}), which may be obtained from Eqs. (31) and (32) by setting $\beta = 0$.

For systems with convergence mode (a), the associated invariant measure may be continuous or have a fractal structure [Karl53]. We shall illustrate this by considering the L_{R-P} algorithm for an automaton with two possible actions. It can be shown that if $\alpha = \beta$, then Eqs. (31) and (32) reduce to an IFS consisting of two mappings, $f_0, f_1 : [0, 1] \to [0, 1]$ and their associated probabilities ϕ_0, ϕ_1, where
$$f_0(x) = \lambda x \qquad f_1(x) = \lambda x + 1 - \lambda \qquad \phi_0 + \phi_1 = 1 \tag{33}$$

Moreover, the probabilities ϕ_0, ϕ_1 become state-independent. We now discuss the behaviour of the invariant measure μ_F of the IFS (Eq. 33) as the parameter λ is varied.

For $\lambda < 1/2$, the support of μ_F is the Cantor set. Indeed, if $\lambda < 1/3$ it is easy to verify that one obtains the usual middle thirds Cantor set.

For $\lambda \geq 1/2$, the support of μ_F is the whole unit interval. It is known [Erdo39] that μ_F is either totally singular or absolutely continuous. Although examples of both cases have been found, e.g., λ_s is singular and λ_c

is absolutely continuous with

$$\lambda_s = \frac{(\sqrt{5}-1)}{2} = \frac{1}{golden\ mean}$$

$$\lambda_c = \sqrt[n]{\frac{1}{2}} \quad n \geq 2$$

not much more is known. It appears, however, that the measure becomes progressively smoother as $\lambda \to 1$. This is illustrated in Figure 1. The unit interval is partitioned into $N = 2100$ subintervals, and the frequency histograms are displayed for parameter values (a) $\lambda = 0.52$, (b) $\lambda = 0.6$ and (c) $\lambda = 0.9$. The probabilities are set to be $\phi_0 = \phi_1 = 1/2$. An example of unequal probabilities is given in Figure 2. It is intriguing that such rich structure can arise from the same family of stochastic processes.

Associative Stochastic Learning Automata

In this section we describe how to construct learning automata which are sensitive to changing contexts within the environment. We follow the neural network approach of Barto et al. [Bart81; Bart85], which we incorporate into the mathematical framework of random IFSs. We define an associative learning automaton as follows:

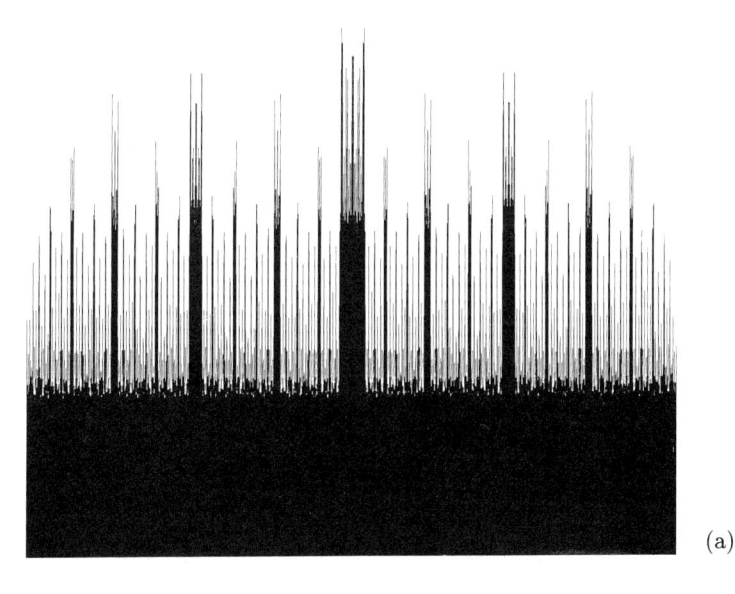

Figure 1. The invariant measure of a random IFS consisting of the two mappings $f_0, f_1 : [0,1] \to [0,1]$ is shown, with $\phi_0 = \phi_1 = 1/2$ and (a) $\lambda = 0.52$.

(b)

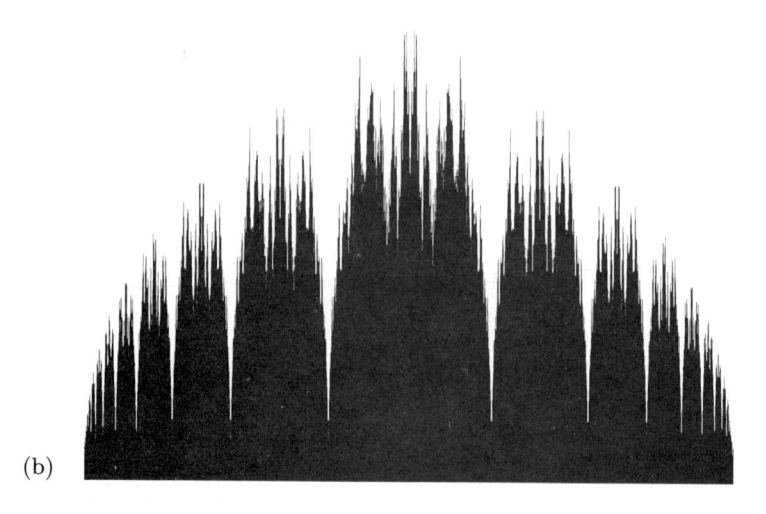

Figure 1 (*Continued*). (b) $\lambda = 0.6$.

(a) At each time-step n the automaton selects an action $y_n \in \{1, 0\}$ according to the dynamical equation

$$y_n = \theta(\boldsymbol{\omega}_n \cdot \mathbf{x}_n + \eta) \tag{34}$$

where θ is the step-function (Eq. 8). In this respect the automaton is

(c)

Figure 1 (*Continued*). (c) $\lambda = 0.9$.

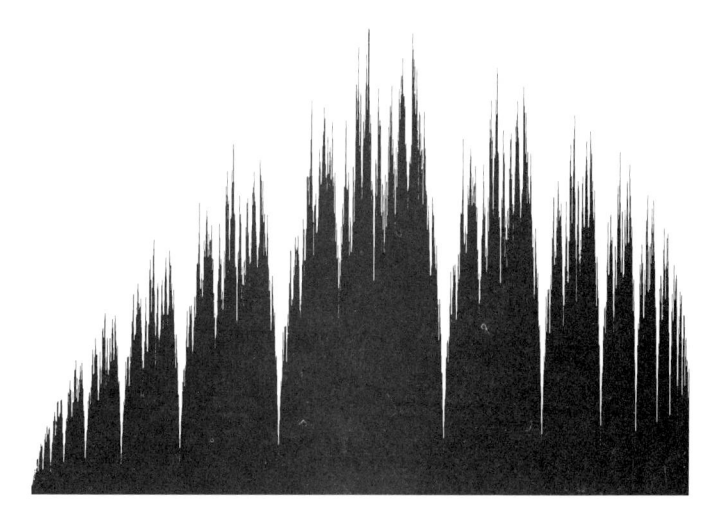

Figure 2. Same as Figure 1 but with parameter values $\phi_0 = 0.6, \phi_1 = 0.4$ and $\lambda = 0.6$.

acting like a binary threshold neuron. The input to the neuron is an m-dimensional vector $\mathbf{x}_n$, which is selected at random from a finite set $X = \{\mathbf{x}^{(1)}, \ldots, \mathbf{x}^{(s)} \in R^m\}$ according to the constant probabilities $\xi^{(a)}, a = 1, \ldots, s$. The input $\mathbf{x}_n$ is interpreted as a context vector reflecting the current state of the environment. The state of the automaton is specified by the weight vector $\mathbf{w}_n \in R^m$; we shall denote weight space by Ω. The sequence of thresholds $\{\eta_n\} \in R$ is a random process with each η_n chosen from the same distribution Ψ. Then, from Eq. (34), Ψ determines the action probabilities associated with a particular context to be

$$p^{(1,a)} \equiv \mathbf{Pr}\{y = 1 \mid \boldsymbol{\omega}, \mathbf{x}^{(a)}\} = 1 - \Psi(-\boldsymbol{\omega}.\mathbf{x}^{(a)}), p^{(0,a)} = 1 - p^{(1,a)} \quad (35)$$

(b) Given a particular context $\mathbf{x}^{(a)} \in X$, the environment sends an evaluative signal $b_n \in \{1, 0\}$ according to the probabilities

$$\pi^{(i,a)} = \mathbf{Pr}\{b_n = 1 \mid y_n = i, \mathbf{x}_n = \mathbf{x}^{(a)}\} \quad (36)$$

The $\pi^{(i,a)}$ are not known *a priori* and are time-independent.

(c) Define an event $\bar{e}_n$ as the triplet

$$\bar{e}_n = (i_n, b_n, a_n) \in \bar{E} \quad (37)$$

where
$$i_n = \begin{cases} 1 & \text{if } y_n = 1 \\ 0 & \text{if } y_n = 0 \end{cases}$$

$$a_n = \begin{cases} 1 & \text{if } \mathbf{x}_n = \mathbf{x}^{(1)} \\ \ldots & \ldots \\ s & \text{if } \mathbf{x}_n = \mathbf{x}^{(s)} \end{cases}$$

and $\bar{E}$ is the set of events $\{(i,b,a) : i,b \in \{1,0\}, 1 \leq a \leq s\}$. Each event now corresponds to the context-dependent pairing of an action with the associated environmental response. To use notation consistent with the previous section, we have distinguished between the index i_n and the action y_n. From (a) and (b) above, given a state $\boldsymbol{w}$, the probability of an event $\bar{e} = (i,b,a)$ is

$$\phi_{\bar{e}}(\boldsymbol{w}) = p^{(i,a)}(\boldsymbol{w}) \xi^{(a)}[\pi^{(i,a)} b + (1 - \pi^{(i,a)})(1-b)] \tag{38}$$

The $\phi_{\bar{e}}$ are probability distributions on the space Ω.

(d) A learning algorithm for the automaton is the set of continuous maps $\{F_{\bar{e}} : \Omega \to \Omega : \bar{e} \in \bar{E}\}$. The dynamical equation for updating the weights is then

$$\boldsymbol{w}_{n+1} = F_{\bar{e}_n}(\boldsymbol{w}_n) \tag{39}$$

where $F_{\bar{e}_n}$ is chosen at random with probability $\phi_{\bar{e}_n}(\boldsymbol{w}_n)$.

We see that the operation of an associative automaton is the updating of the weights of a binary neuron on the basis of context-dependent responses from the environment. Ideally, one wants the automaton eventually to respond to each context vector $\mathbf{x}^{(a)}$, with the action $y = 1$ if $\pi^{(1,a)} > \pi^{(0,a)}$, and $y = 0$ if $\pi^{(1,a)} < \pi^{(0,a)}$. To discuss the performance of the automaton, we note that the set $\{(F_{\bar{e}}, \phi_{\bar{e}}) : \bar{e} \in \bar{E}\}$ defines a random IFS on Ω. Following the previous section, we introduce the stochastic kernel

$$K(\boldsymbol{w}, B) = \sum_{\bar{e} \in \bar{E}} \phi_{\bar{e}}(\boldsymbol{w}) \mathcal{X}_B(F_{\bar{e}}(\boldsymbol{w})) \qquad B \in \mathbf{B}(\Omega) \tag{40}$$

which induces a sequence of measures

$$\mu_{n+1} = \int_{\Omega} K(\boldsymbol{w}, .) d\mu_n(\boldsymbol{w}) \tag{41}$$

The measure μ_n may be used to define the probability

$$q_n^{(i,a)} = \int_{\Omega} p^{(i,a)}(\boldsymbol{w}) d\mu_n(\boldsymbol{w}) \tag{42}$$

It is important to distinguish between $q_n^{(i,a)}$ and $p^{(i,a)}(\boldsymbol{w})$ in Eq. (42). The latter is the probability that an automaton in a particular state $\boldsymbol{w}$ has action $y = i$ given a context vector $\mathbf{x}^{(a)}$ and is appropriate when considering dynamics. The former, on the other hand, is the probability that an element of an ensemble of automata, with distribution $\mu_n \in \mathbf{P}(\Omega)$, has action $y = i$ given $\mathbf{x}^{(a)}$ and is appropriate when considering performance.

Following Eq. (23), the probability of success at step n for an associative learning automaton is

$$Q_n = \sum_{i=1}^{2} \sum_{a=1}^{s} \xi^{(a)} \pi^{(i,a)} p^{(i,a)}(\boldsymbol{w}_n) \tag{43}$$

It is maximised when the optimal action for each input context vector occurs with probability one, in which case

$$Q_{\max} = \sum_{a=1}^{s} \xi^{(a)} \max\left\{\pi^{(1,a)}, \pi^{(0,a)}\right\} \qquad (44)$$

If, on the other hand, the actions 1 and 0 are chosen with equal probability, the probability of success is

$$Q^0 = \sum_{i=1}^{2} \sum_{a=1}^{s} \frac{\xi^{(a)} \pi^{(i,a)}}{2}$$

Assuming that the sequence μ_n converges in distribution, we say that an associative learning automaton is

- expedient if $\lim_{n\to\infty} \mathbf{E}(Q_n) > Q^0$;
- optimal if $\lim_{n\to\infty} \mathbf{E}(Q_n) = Q_{\max}$;
- ϵ-optimal if $\lim_{n\to\infty} \mathbf{E}(Q_n) > Q_{\max} - \epsilon$ can be obtained for any $\epsilon > 0$ by a suitable choice of learning algorithm parameters.

Here

$$\mathbf{E}(Q_n) = \sum_{i=1}^{2} \sum_{a=1}^{s} \xi^{(a)} \pi^{(i,a)} q_n^{(i,a)} \qquad (45)$$

where $q_n^{(i,a)}$ satisfies Eq. (42).

We remark:

A class of learning algorithms has been found of the general form [Bres89b]

$$\boldsymbol{\omega}_{n+1} = \boldsymbol{\omega}_n + f_{e_n}^{(a_n)}(\boldsymbol{\omega}_n.\mathbf{x}^{(a_n)})\lambda^{(a_n)} \qquad (46)$$

where $\lambda^{(a)}$ are the dual vectors satisfying

$$\lambda^{(a)}.\mathbf{x}^{(b)} = \delta_{ab}$$

(We are assuming that the $\mathbf{x}^{(a)}$ are linearly independent.) The probability of choosing $f_e^{(a)}$ is given by Eq. (38), with $\bar{e} = (e, a)$. The essential feature of these algorithms is that the distribution Ψ is defined such that Eq. (35) induces a homeomorphism between the space of action probabilities and the space of components $\boldsymbol{\omega}.\mathbf{x}^{(a)}$ for all $a = 1, \ldots, s$. Then, the convergence properties of Eq. (46) may be analysed along similar lines to standard automata (see the preceding section on stochastic learning automata); and both expedient and ϵ-optimal schemes can be obtained by appropriate choices of the functions $f_e^{(a)}$.

The above formalism may be extended to a single-layered feed-forward network, in which each binary neuron receives the same input context vector $\mathbf{x}$ and there are no connections within the layer. If there are N neurons, then there are 2^N possible output actions. The probability distributions associated with IFS-type learning algorithms will be Nth-order polynomials in action probabilities, rather than linear as in Eq. (37). Extensions to networks of interacting neurons would need to take into account the fact that, in general, the input to any neuron will consist of signals from other neurons, in addition to external environmental signals.

The random IFS formalism includes learning rules that lead to an invariant measure on weight space with a fractal structure. This is the analogue of the fractals found by Karlin [Karl53] in his analysis of standard learning automata, examples of which are given in Figures 1 and 2. The possibility of fractals, a natural consequence of our construction, does not appear to have been realised in previous treatments of adaptive neural networks. It remains an open question whether such fractals have any significance in neuronal information processing.

It is important to understand the role of threshold noise in associative learning automata and to make a comparison of the different possible forms for the distribution Ψ. Following certain stochastic models of neuronal processing [Bres89a; Bres90], we note that there is a more general approach to incorporating noise into associative learning automata. We replace Eq. (34) with

$$y_n = \theta\left(\sum_{I=1}^{m} \omega_n^I \kappa_n^I x_n^I + \eta_n\right) \qquad (47)$$

where $\{\omega_n^I : I = 1, \ldots, m\}$ are the components of $\boldsymbol{\omega}_n$, etc. The random variables $\kappa_n \in R^m$ and $\eta_n \in R$ are noise terms. That is, the sequence $\{(\kappa_n, \eta_n)\}$ is a random process, with each $\{(\kappa_n, \eta_n)\}$ having the same time-independent distribution Ψ. Integrating Eq. (47) over Ψ leads to the generalised action probabilities

$$p^{(1,a)}(\boldsymbol{\omega}) = \int \theta\left(\sum_{I=1}^{m} \omega^I \kappa^I x^{(a)I} + \eta\right) d\Psi(\kappa, \eta) \qquad p^{(0,a)} = 1 - p^{(1,a)} \quad (48)$$

which then may be substituted into Eq. (38). The specific form of Eq. (47) is based on a model of synaptic processes in biological neurons; a typical distribution for κ would be a Poisson or binomial distribution [Bres90]. Note that the learning algorithm (Eq. 46) relies on the probabilities $p^{(i,a)}$ being functions of the inner product $\boldsymbol{\omega}.\mathbf{x}^{(a)}$. This functional form will, in general, no longer hold when there is noise associated with the weights as in Eq. (47).

These issues will be considered in more detail elsewhere (see [Bres89b]).

Conclusions

In this paper we have underlined certain connections between Iterated Function Systems and neural networks and illustrated this in the context of image generation and stochastic learning automata. Neural networks provide a means of parallel information processing which may have uses in data compression. On the other hand, Iterated Function Systems provide a natural and broad framework for studying neural networks which we have used to construct associative learning automata.

Acknowledgements. J. Stark would like to thank Professor E.C. Zeeman for first drawing his attention to the examples shown in Figure 1.

REFERENCES

[Barn85]
  Barnsley, M.F., Demko, S., Elton, J., and Geronimo, J., Markov processes arising from iteration with place-dependent probabilities, *Pre-print*, Georgia Inst. Technology, 1985.

[Barn88a]
  Barnsley, M.F., *Fractals Everywhere*, San Diego: Academic Press, 1988.

[Barn88b]
  Barnsley, M.F., and Sloan, A.D., A better way to compress images, *Byte*, Vol. 13, pp. 215–223, 1988.

[Bart81]
  Barto, A.G., Sutton, R.S., and Brouwer, P.S., Associative search network, *Biol. Cybern.*, Vol. 40, pp. 201–211, 1981.

[Bart85]
  Barto, A.G., and Anandan, P., Pattern-recognising stochastic learning automata, *IEEE Trans. Syst. Man and Cybern.*, Vol. 15, pp. 360–375, 1985.

[Bres89a]
  Bressloff, P.C., Neural networks and random iterative maps, in *Neural Computation*, Mannion, C., and Taylor, J.G., Eds., pp. 155–162, Bristol, UK: Adam Hilger, 1989.

[Bres89b]
  Bressloff, P.C., and Stark, J., Associative reinforcement learning based on iterated function systems, submitted to *IEEE Trans. Syst. Man and Cybern.*, 1989.

[Bres90]
  Bressloff, P.C., and Taylor, J.G., Random iterative networks, *Phys. Rev. A*, Vol. 41, pp. 1126–1137, 1990.

[Elto87]
  Elton, J.H., An ergodic theorem for iterated maps, *Ergod. Th. and Dynam. Sys.*, Vol. 7, pp. 481–488, 1987.

[Erdo39]
  Erdos, P., On a family of symmetric Bernouilli convolutions, *Amer. Jour. Math.*, Vol. 61, pp. 974–976, 1939.

[Fede69]
  Federer, H., *Geometric Measure Theory*, New York: Springer-Verlag, 1969.

[Falc85]
Falconer, K.J., *The Geometry of Fractal Sets*, Cambridge, UK: Cambridge Univ. Press, 1985.

[Hutc81]
Hutchinson, J.F., Fractals and self-similarity, *Indiana Univ. Jour. of Math.*, Vol. 30, pp. 713–747, 1981.

[Karl53]
Karlin, S., Some random walks arising in learning models, *Pacific Jour. Math.*, Vol. 3, pp. 725–756, 1953.

[Laks81]
Lakshmivarahan, S., *Learning Algorithms Theory and Applications*, New York: Springer-Verlag, 1981.

[Nare74]
Narendra, K.S., and Thathachar, M.A.L., Learning automata—A survey, *IEEE Trans. Syst. Man and Cybern.*, Vol. 4, pp. 323–334, 1974.

[Norm68]
Norman, M.F., Some convergence theorems for stochastic learning models with distance diminishing operators, *Jour. Math. Psychol.*, Vol. 5, pp. 61–101, 1968.

[Norm72]
Norman, M.F., *Markov Processes and Learning Models*, New York: Academic Press, 1972.

[Prus88]
Prusinkiewicz, P., and Sandness, G., Koch curves as attractors and repellors, *IEEE Comput. Graph. and Appl.*, Vol. 8, pp. 26–40, 1988.

[Redd88]
Reddaway, S.F., Fractal graphics and image compression on a DAP, in *International Specialist Seminar on the Design and Application of Parallel Digital Processors* (IEE Conf. Publ. 298), p. 201, 1988.

[Sene81]
Seneta, W., *Non-negative Matrices and Markov Chains*, New York: Springer-Verlag, 1981.

[Star89]
Stark, J., Iterated function systems as neural networks, submitted to *Neural Networks*, 1989.

Colour Plates

Plate 1. A three-dimensional fractal tree created by Hugh Mallinder.

Plate 2. A fractal landscape incorporating terrain, sea, clouds and mist; created by Semannia Luk-Cheung.

Plate 3. A fractal landscape incorporating terrain, sea, clouds and mist; created by Semannia Luk-Cheung.

Plate 4. The Mandelbrot set M_3. p and q both run from -1.5 to 1.5, with 252 points in each direction.

Plate 5. A DLA simulation of 10,000 particles on a 500×500 lattice.

Plate 6. A DBM simulation of 4157 particles on a 300 × 300 lattice.

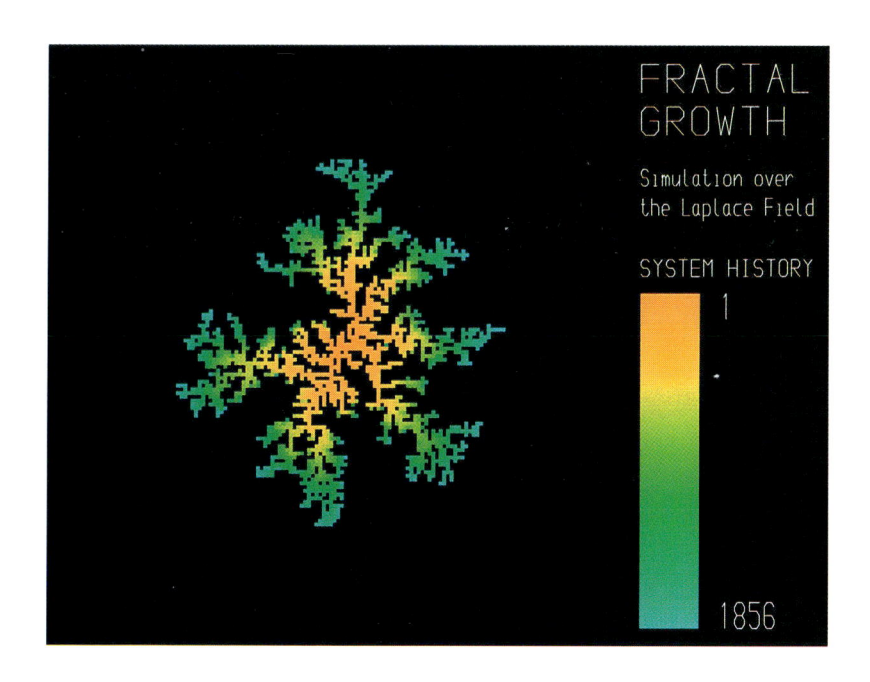

Plate 7. The DBM baseline simulation with $\mu = 1$.

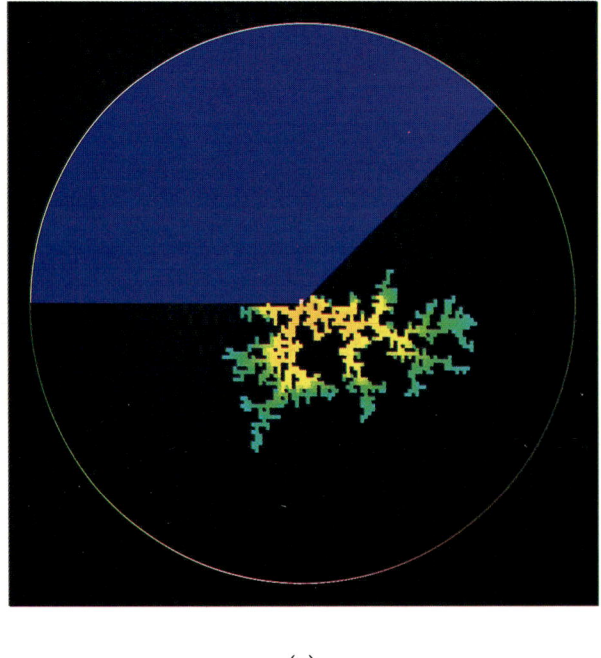

Plate 8. DBM simulations constrained to (a) $\theta = 1.25\pi$ and (b) $\theta = 0.5\pi$.

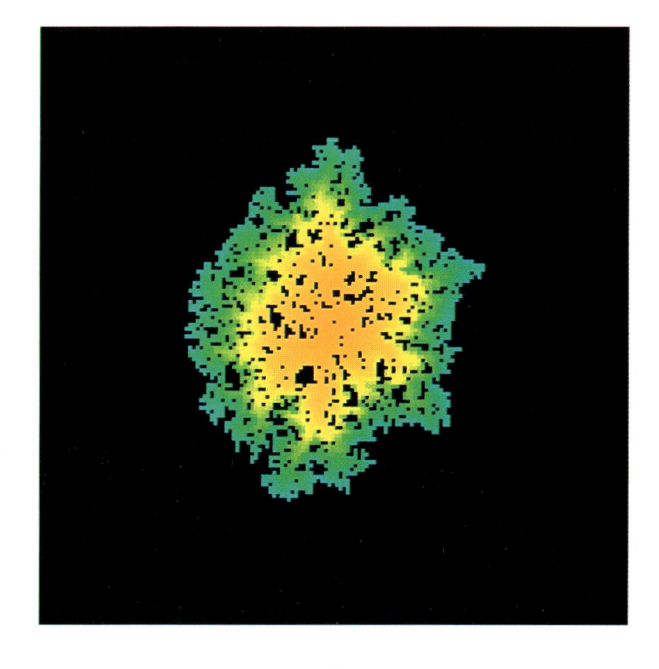

(a)

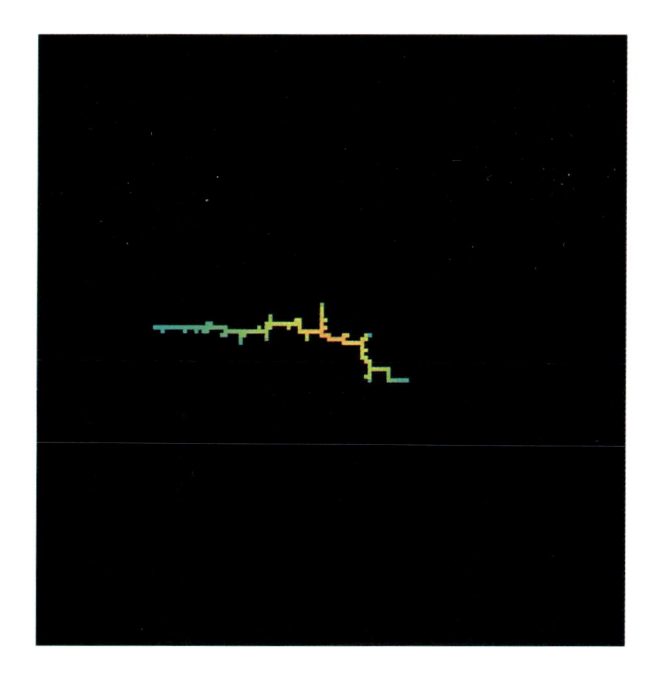

(b)

Plate 9. DBM simulations with (a) $\mu = 0.25$ and (b) $\mu = 4$.

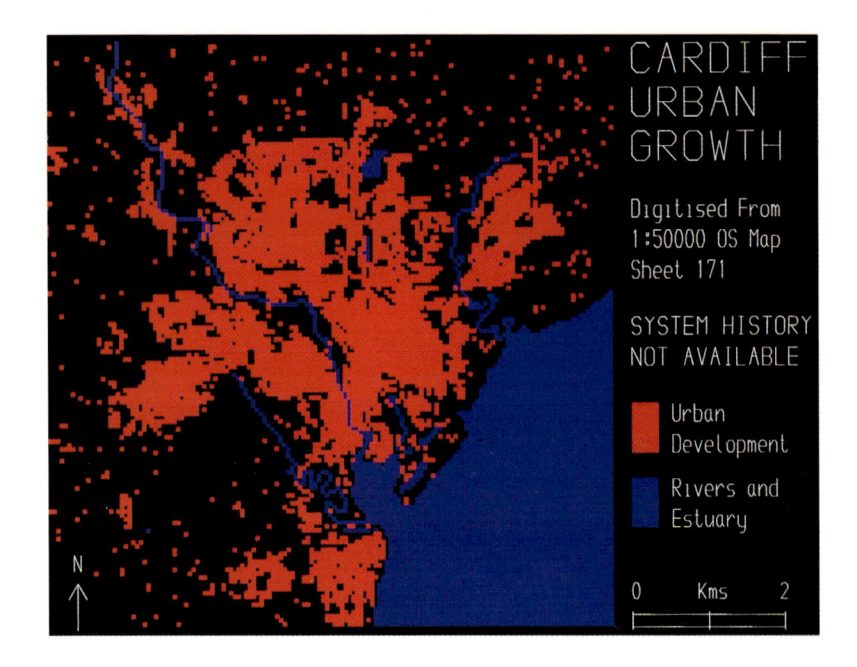

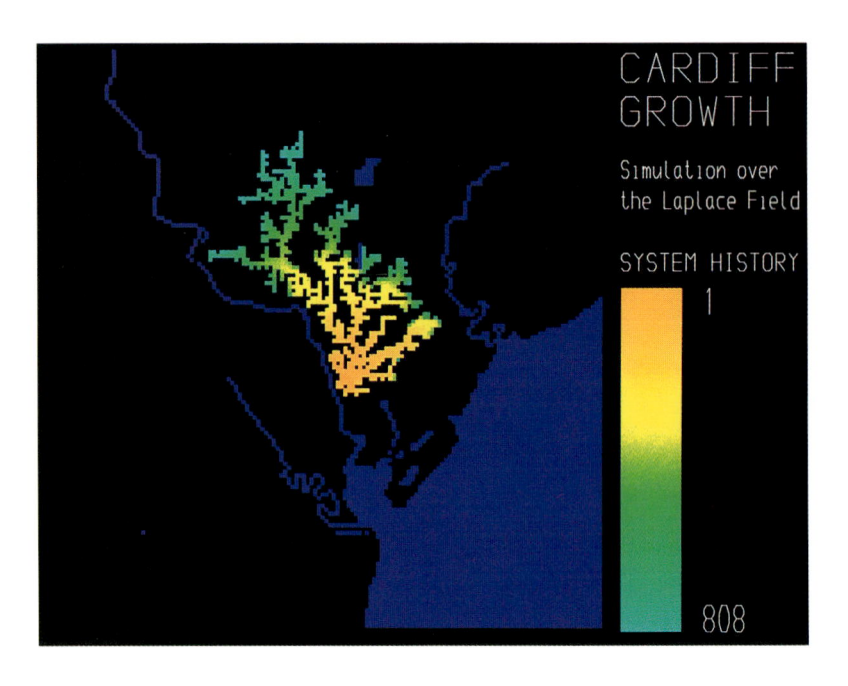

Plate 10. (a) The urban morphology of Cardiff; (b) the DBM baseline simulation of Cardiff.

(a)

(b)

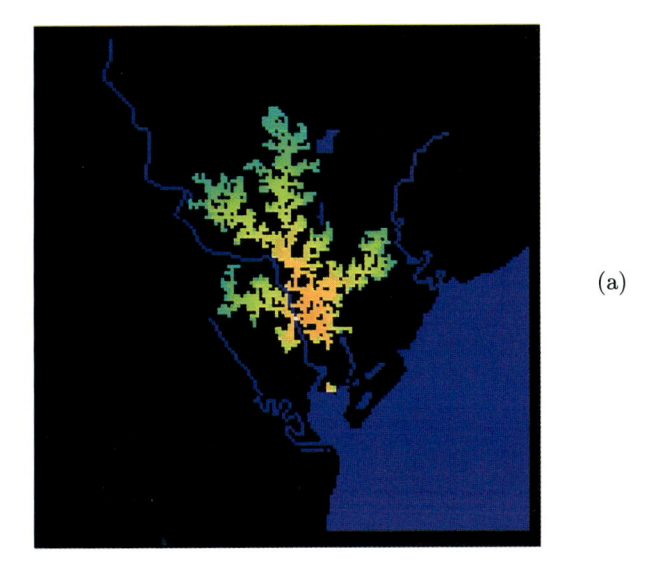

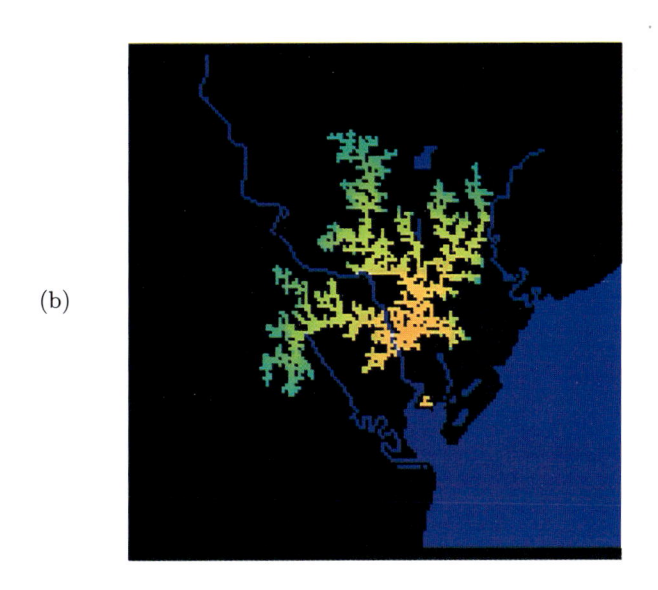

Plate 11. DBM Simulations of Cardiff with (a) $\mu = 1$; (b) $\mu = 0.75$; (c) $\mu = 0.5$; (d) $\mu = 0.25$; (e) $\mu = 0.01$.

(c)

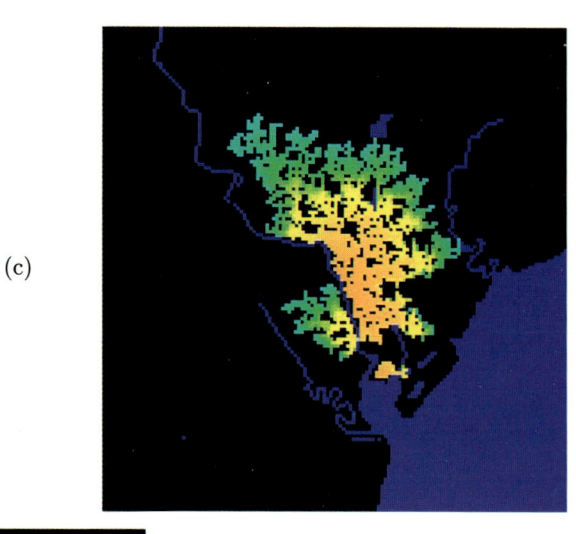

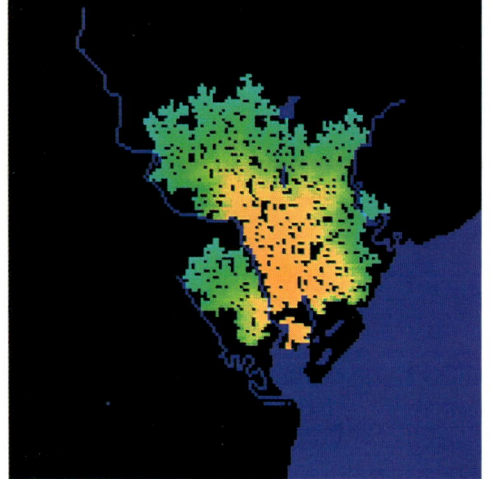

(d)

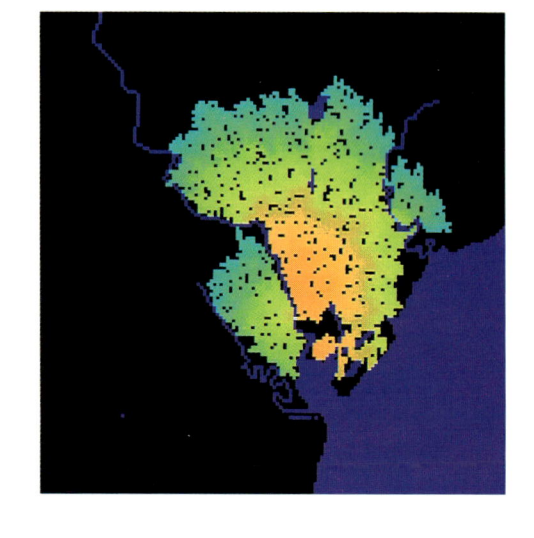

(e)

Plate 11. (cont.)

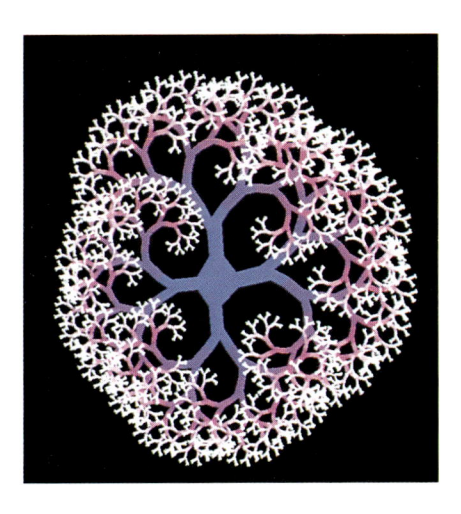

Plate 12. Ramifying fractal resulting from the geometric production rule shown in Figure 2 (see p. 73).

Plate 13. Irregular ramifying fractal resulting from the geometric production rule shown in Figure 3 (see p. 74).

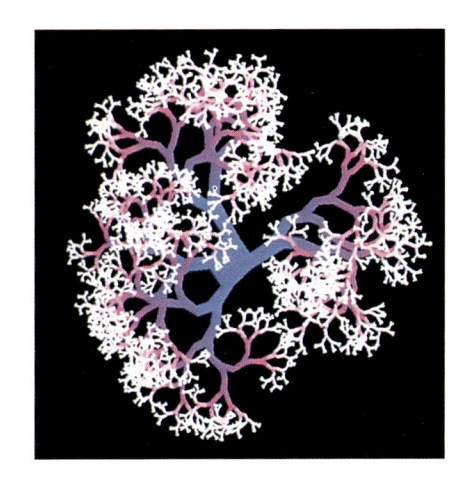

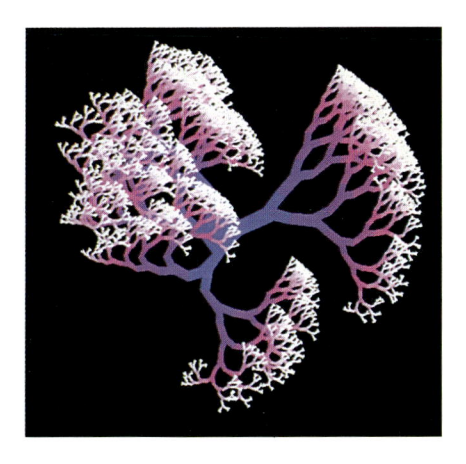

Plate 14. Irregular ramifying fractal that is growing towards a prevailing direction. The fractal results from the geometric production rule shown in Figure 3 (see p. 74).

Plate 15. Irregular ramifying fractal, in which intersecting branches of the fractal approximant shown in Plate 14 are removed by a postprocessing function.

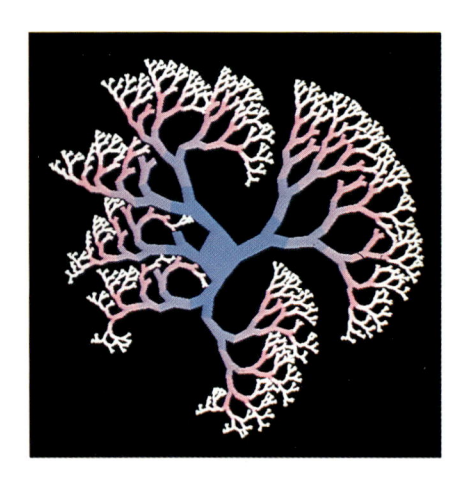

Plate 16. Irregular ramifying fractal, where all fertile sites are generating a maximum of *nretry* branches, until a branch is found that is not intersecting the fractal approximant (see Figure 5, p. 76).

Plate 17. Irregular ramifying fractal, where all fertile sites are generating a maximum of *nretry* branches, until a branch is found that is not intersecting the fractal approximant (see Figure 5, p. 76) or the box surrounding the object.

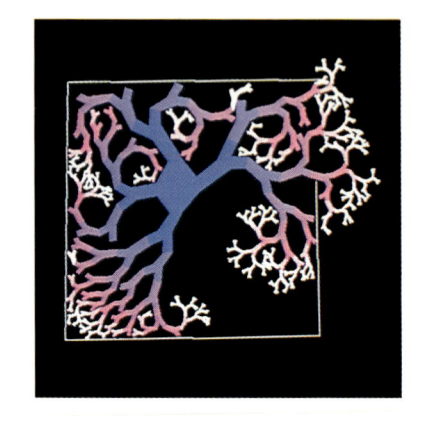

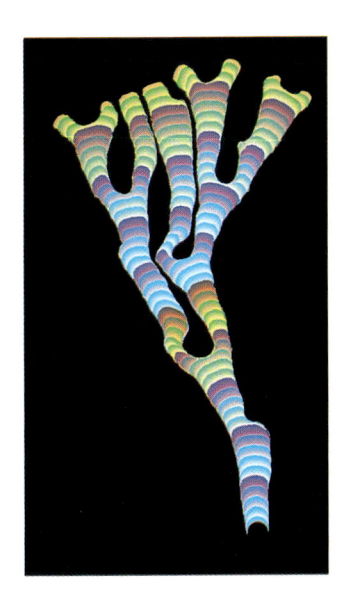

Plate 18. An object generated with the model of Figure 9 (form G, see p. 80). It is constructed with the value $n = 18$ in $f(\alpha)$ in Eq. (6) (see p. 84) in 220 iterations.

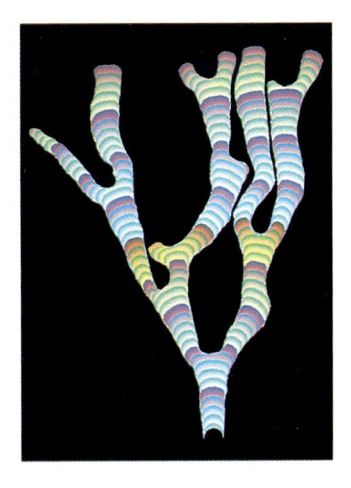

Plate 19. An object generated with the model of Figure 9 (form I), constructed with the value $n = 18$ in $f(\alpha)$ in Eq. (7) (see p. 85) in 220 iterations. Here the growth process was not disturbed by the function $g()$ ($lowest\_value = 1.0$).

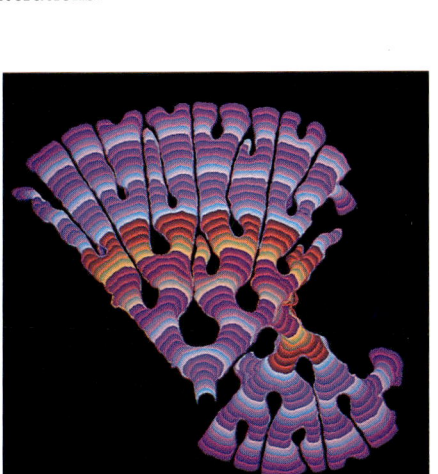

Plate 20. An object generated with the model of Figure 9 (form H). It is constructed with the value $n = 18$ in $f(\alpha)$ in Eq. (6) and by disturbing the growth process by multiplying the product $f(\alpha) \cdot h(rad\_curv)$ in Eq. (6) with the function $g()$ (see Eq. 4, p. 82).

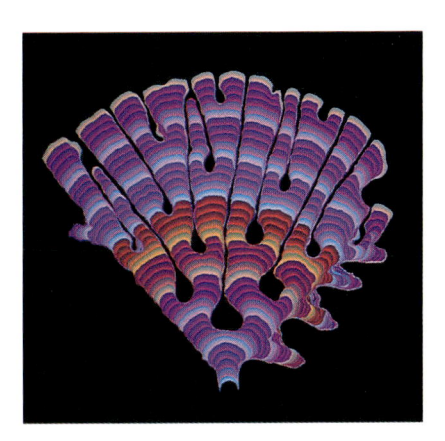

Plate 21. An object generated with the model of Figure 9 (form I). It is constructed with the value $n = 18$ in $f(\alpha)$ in Eq. (7) and by disturbing the growth process by multiplying the product $f(\alpha) \cdot h(rad\_curv) \cdot k(\beta)$ in Eq. (7) with the function $g()$ (see Eq. 4).

Plate 22. 'Spirit Lake'. This image by F.K. Musgrave shows the effect of modulating amplitudes of summed terms in the rescale-and-add method. Texture mapping is effectively used to increase visible surface complexity. The physical rainbow model is from Musgrave [Musg89b]. ©F.K. Musgrave and B.B. Mandelbrot.

Plate 23. 'Lethe'. Dramatic effects in this ray-traced scene by F.K. Musgrave are achieved by steep fractal mountains and surrealistic lighting. Note the procedural texture map simulating sedimentary rock strata. ©F.K. Musgrave and B.B. Mandelbrot.

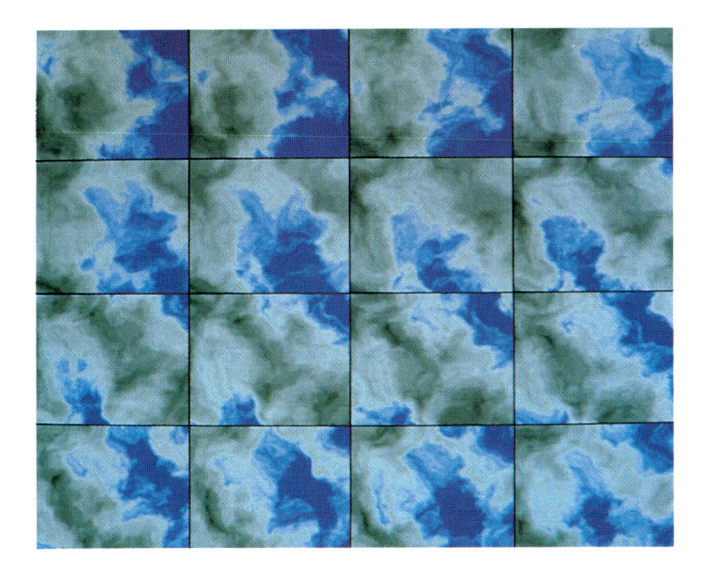

Plate 24. 16 frames from an animation of clouds. This is an application of random fractals in three variables.

Plate 25. Fractal planet based on the rescale-and-add method. The surface properties on the sphere depend on the location. The fractal dimension of the coastlines near the equator is equal to 1.2, whereas at the poles it is 1.9. The colouring depends also on the latitudes such that 'polar caps' and some 'mountains' are covered by 'ice'.

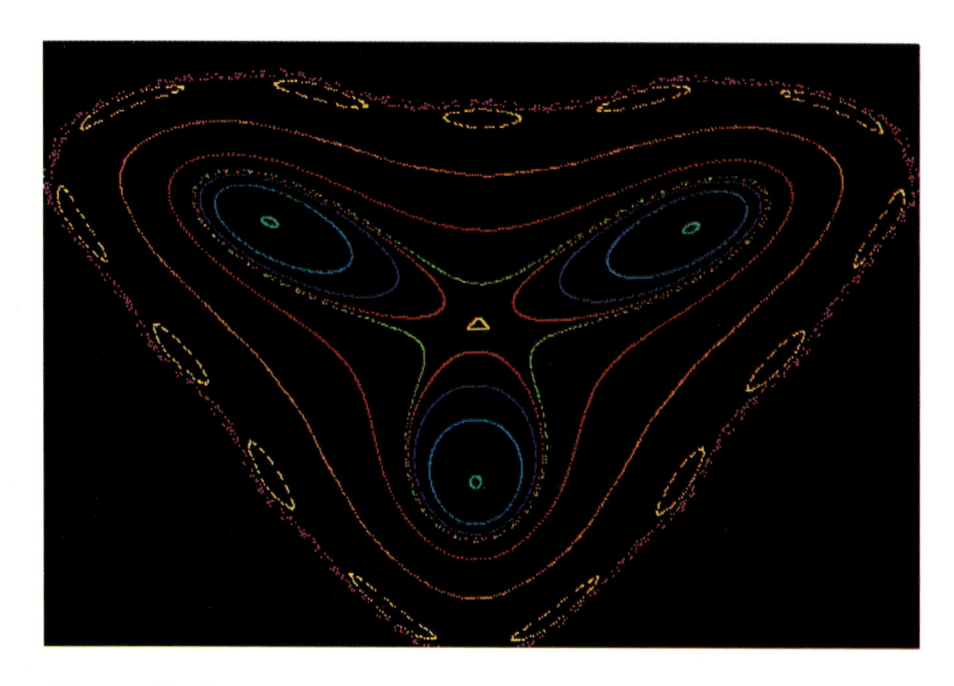

Plate 26. The Poincaré map for different initial values of the momentum.
©M.M. Novak 1989.

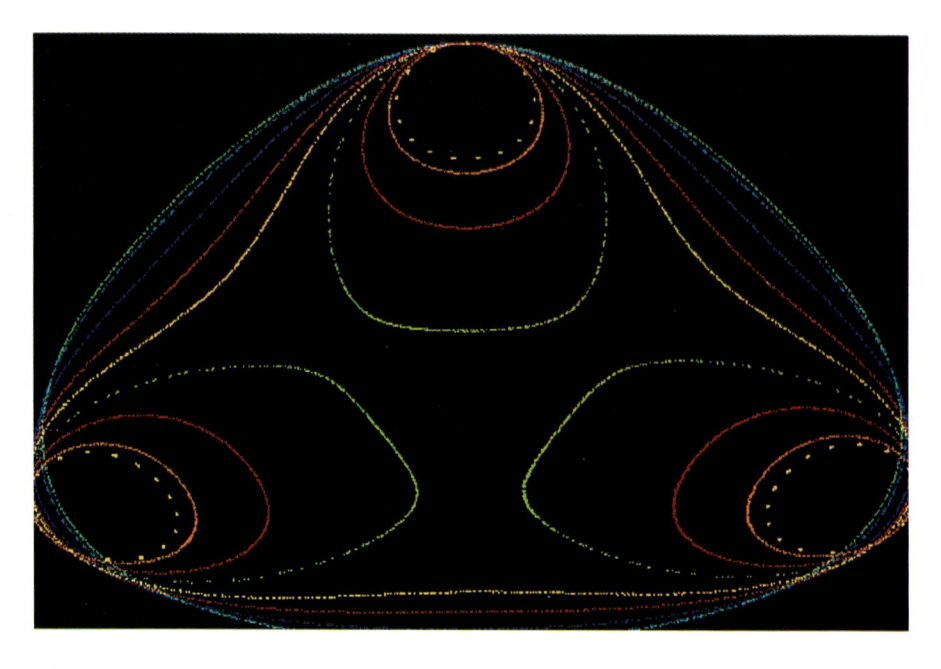

Plate 27. The Poincaré map for different values of the magnetic field.
©M.M. Novak 1989.

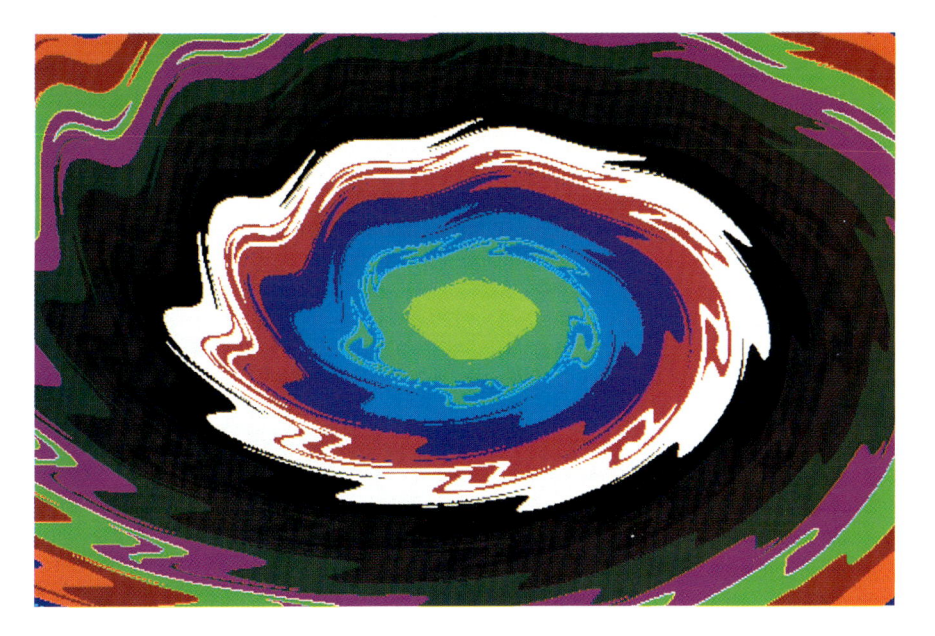

Plate 28. The Hamiltonian function resulting when the perturbation parameter is set to 0.05; initial values are in the (x, p) space. ©M.M. Novak 1989.

Plate 29. The Hamiltonian function resulting when the perturbation parameter is set to 0.2; initial values are in the (x, p) space. ©M.M. Novak 1989.

Plate 30. The Hamiltonian function showing magnified central region, for an intermediate value of the perturbation parameter; initial values are in the (x, p) space. ©M.M. Novak 1989.

Plate 31. The Hamiltonian function showing magnified central region, for an intermediate value of the perturbation parameter; the frequency of the perturbation has been increased. Initial values are in the (x, p) space. ©M.M. Novak 1989.

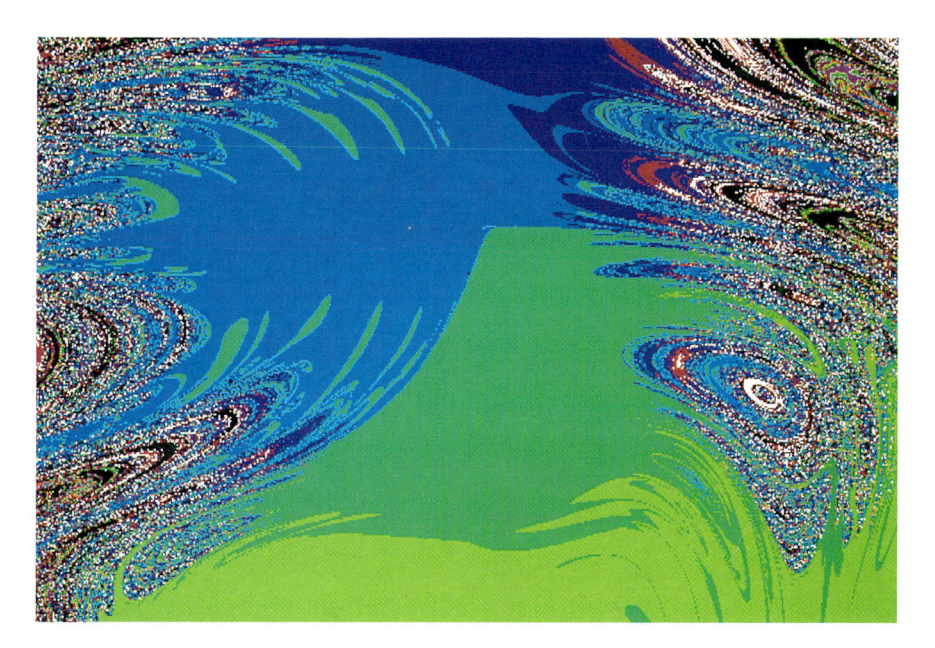

Plate 32. The Hamiltonian function for different values of the perturbation parameters and momenta; parameter space. ©M.M. Novak 1989.

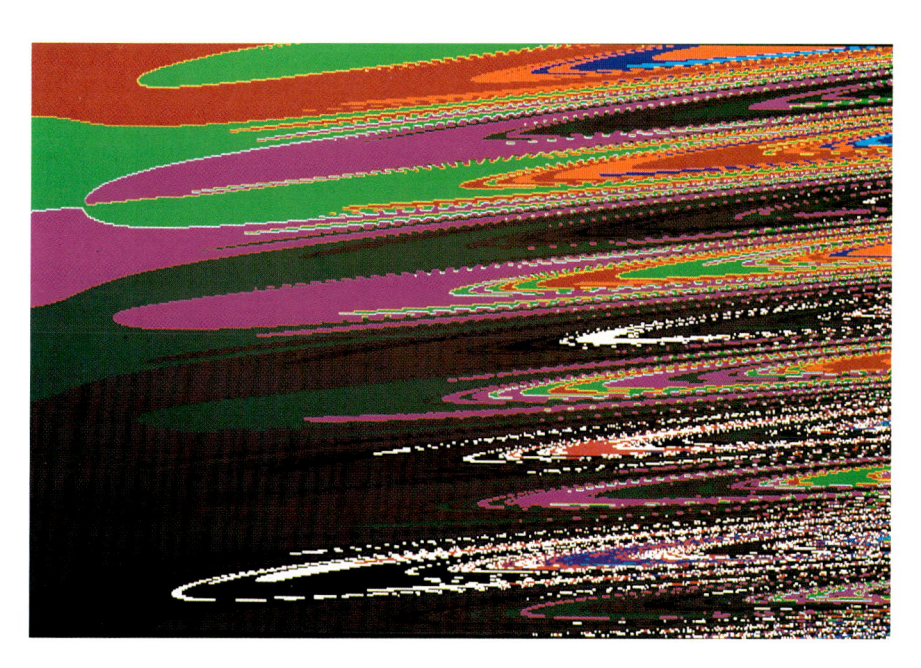

Plate 33. The Hamiltonian function for positive values of the perturbation, in the relativistic velocities domain; parameter space. ©M.M. Novak 1989.

Plate 34. The Hamiltonian function for the values of the perturbation parameter centred around the origin, restricted relativistic domain; parameter space.
©M.M. Novak 1989.

Plate 35. The Hamiltonian function showing a magnified region given in Plate 32; parameter space. ©M.M. Novak 1989.

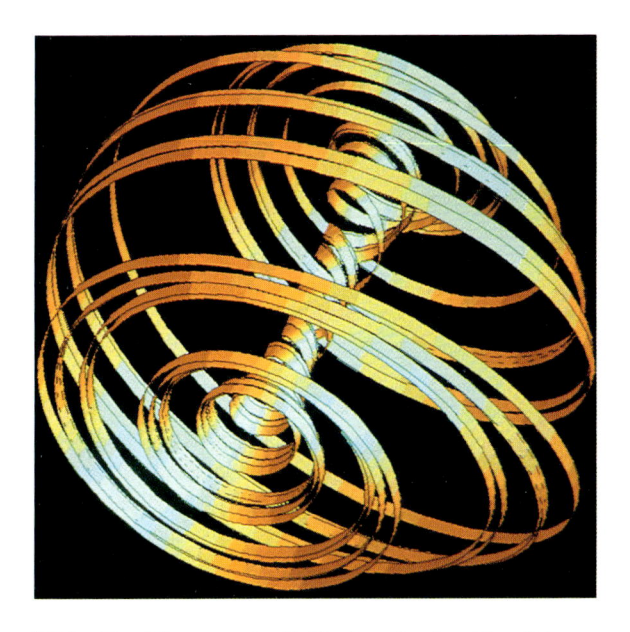

Plate 36. Phase portrait displayed using the ribbon method.

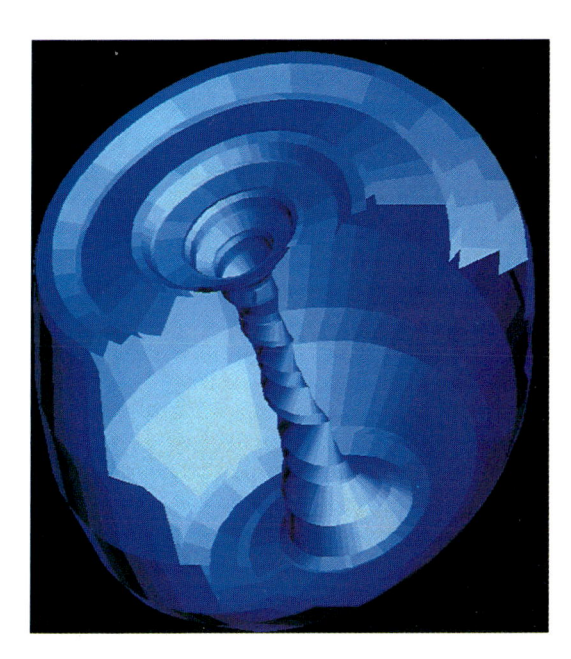

Plate 37. Phase portrait displayed using the surface method (with cutaway section).

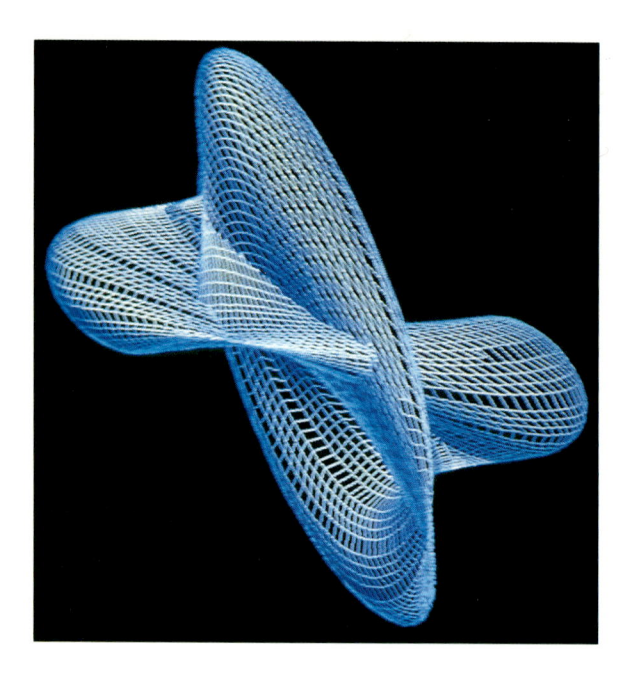

Plate 38. Five-dimensional phase portrait with an apparent self-intersection when projected onto three dimensions.

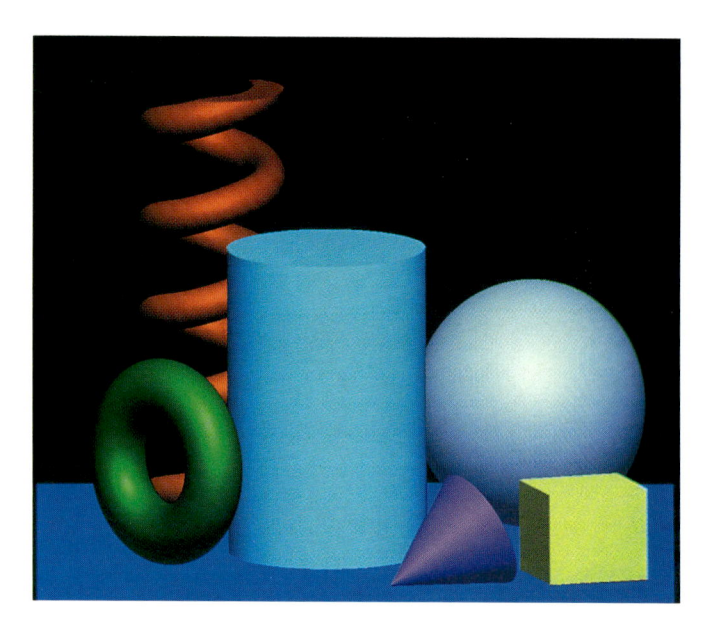

Plate 39. Some of the more commonly used WINSOM primitive shapes.

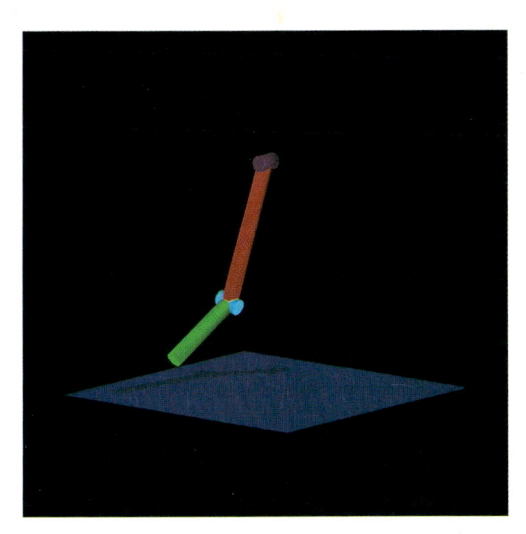

Plate 40. A snapshot of the motion of the excited coupled pendula.

Plate 41. The two distinct attractors (using coodinate choice A—see text, p. 235). The surface has been coloured according to the sign of the 'missing' angular velocity.

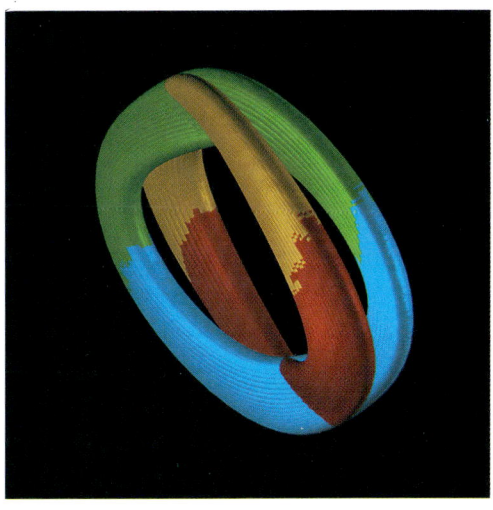

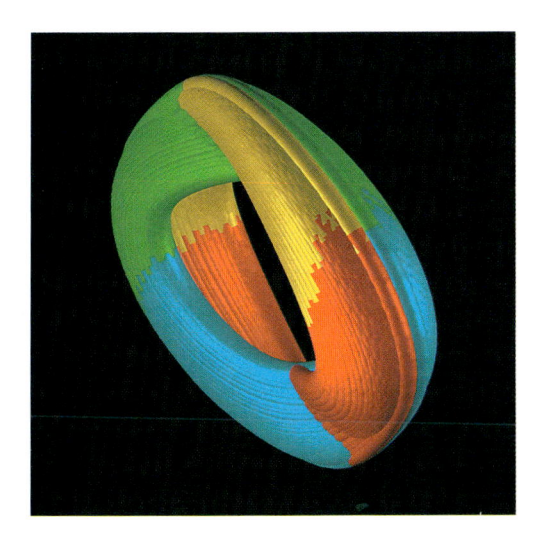

Plate 42. The two attractors just before they join (using coordinate choice A—see text, p. 235). The surface has been coloured according to the sign of the 'missing' angular velocity.

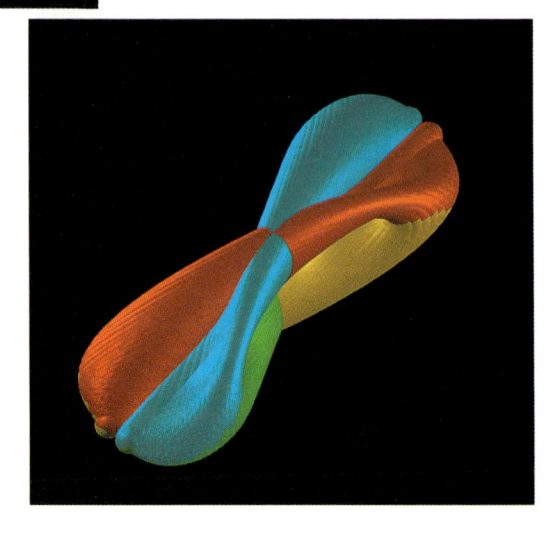

Plate 43. The attractor that results from the gluing bifurcation (using coordinate choice *A*—see text, p. 235). The surface has been coloured according to the sign of the 'missing' angular velocity.

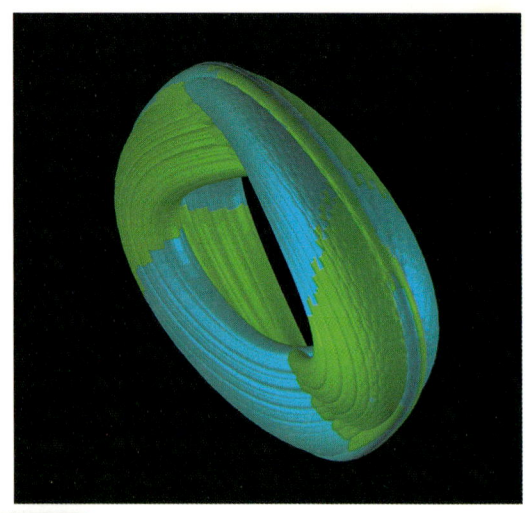

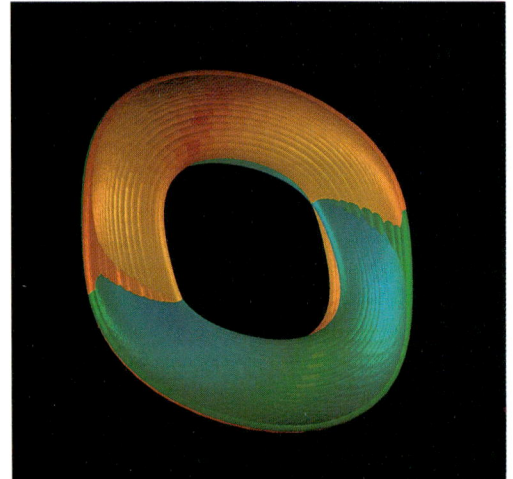

Plate 44. The two attractors just before they join (using coodinate choice *A*—see text, p. 235). The surface has been coloured according to the magnitude of the 'missing' angular velocity.

Plate 45. The two attractors just before they join (using coordinate choice *B*—see text, p. 235). The surface has been coloured according to the sign of the 'missing' angular velocity.

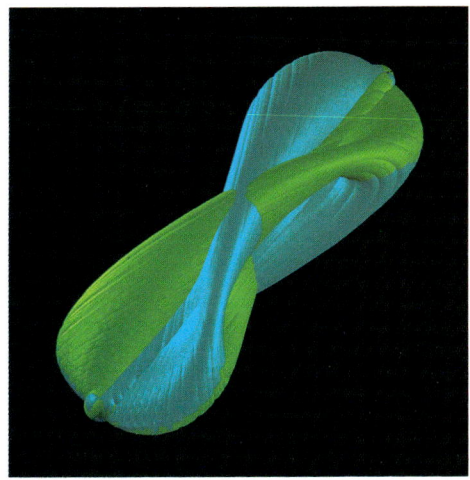

Plate 46. The attractor that results from the gluing bifurcation (using coordinate choice B—see text, p. 235). The surface has been coloured according to the sign of the 'missing' angular velocity.

Plate 47. The two attractors just before they join (using coodinate choice A—see text, p. 235). The surface has been coloured according to a texture mapping that involves both 'missing' variables.

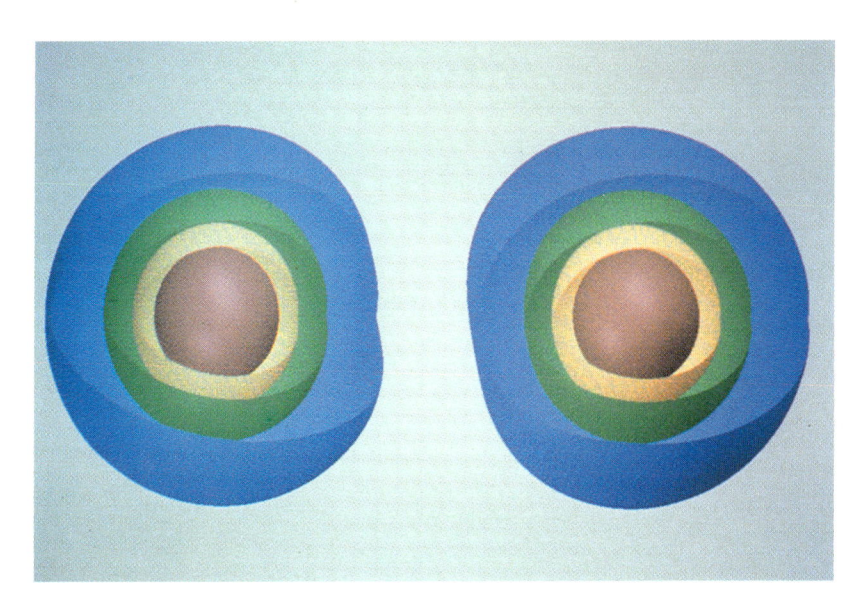

Plate 48. Four energy contour surfaces for two static monopoles at separation 4/3 units.

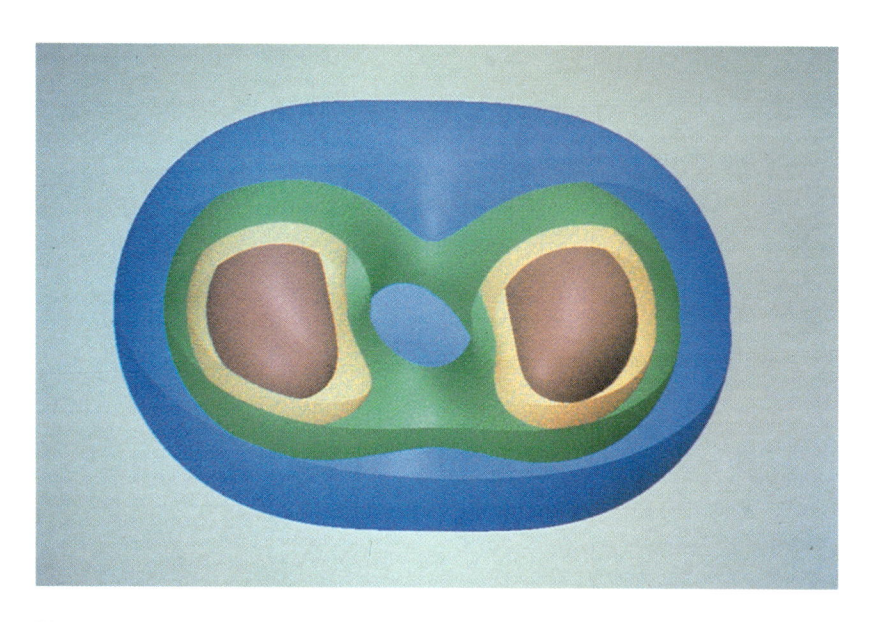

Plate 49. Four energy contour surfaces for two static monopoles at separation 2/3 units.

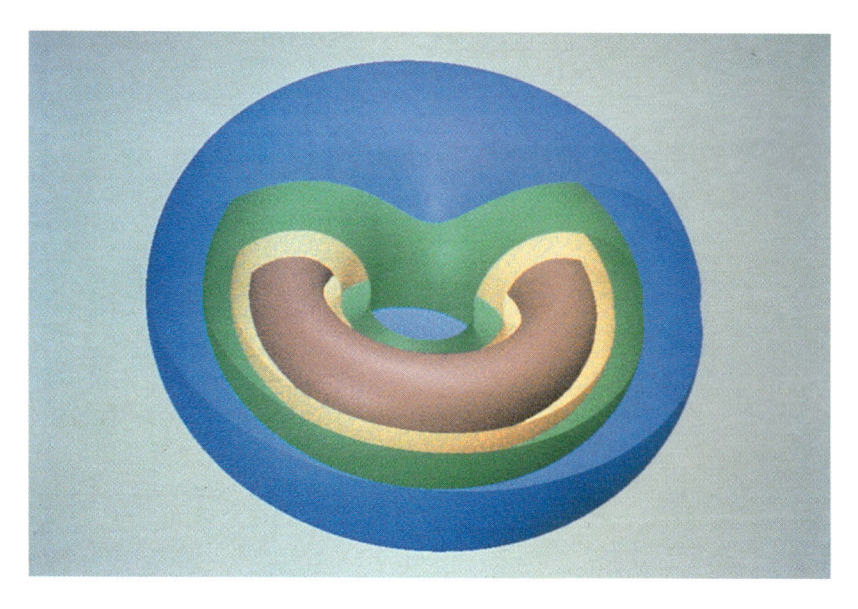

Plate 50. Four energy contour surfaces for two static monopoles at zero separation.

2 Chaos

The Roots of Chaos—A Brief Guide

Tony Crilly

Abstract

> It has been known since the time of Poincaré that simple deterministic systems can give rise to unpredictable behaviour. This chapter reviews some of the leading ideas of chaos and outlines some of its applications which have been made during the 1980s. It is intended to serve as background support for the papers presented in this volume. No attempt has been made to be encyclopedic in what has become a vast enterprise. A list of general references is appended.

> "I resolved to shun story telling. I would write about life. ... Nothing would be left out. Let others bring order to chaos. I would bring chaos to order, instead."—Kurt Vonnegut (*Breakfast of Champions*)

Introduction

The rise of chaos has been meteoric. During the 1980s it received widespread prominence, and some scientists have even placed it alongside two other great revolutions of physical theory in the twentieth century—relativity and quantum mechanics. While those theories challenged the Newtonian system of dynamics, chaos has questioned traditional beliefs from within the Newtonian framework.

The growth of chaos, the fascinating combination of order and disorder brought about by continual instability, is linked with the rapid development of powerful computers. Within the last fifteen years, successful innovation in computer graphics is one of the factors which has enabled scientists and mathematicians to make progress in nonlinear systems from which chaos is derived.

Through realistic computer simulations and video techniques, scientists can now view the evolution of dynamical systems and the complex chaotic effects of the underlying differential equations. Chaos has gained wide appeal, and attempts have been made to identify the phenomenon in many different branches of science. These include traditional fields such as biology, physics, chemistry, mathematics, astronomy, economics and geography. Applications which form the subject matter of this book are largely centred on the physical sciences, but there is also a paper [Lans91] which argues that chaos can be used to investigate and explore the field of art and design.

The Historical Background—Linear Systems

In the physical sciences, the achievements of scientists and mathematicians of the eighteenth and nineteenth centuries were largely bound up with modelling natural phenomena using special differential equations. These could be solved exactly in terms of a function given by a formula—the closed form solution; this led to a study of differential equations with the rather restricted view of finding solutions of this kind.

A broader approach was to classify differential equations by type, and the greatest effort was extended on linear equations. It was a natural approach to take, for the mathematics of linear equations is at least tractable. Using scientific principles such as Newton's laws of motion, single linear differential equations, or sets of linear differential equations, were deduced which could then be solved once the system's initial conditions were known. Broadly speaking, in a linear physical system outputs are proportional to inputs, and linear differential equations are correspondingly straightforward. They have pleasant mathematical properties. For example, two individual solutions of a linear differential equation can be added to produce a third. Using this additive property, a more general solution of the equation can be constructed by superimposing the basic solutions.

Among the linear differential equations dealt with during the earlier part of the last century were the classical wave equation and the heat conduction equation associated with such illustrious figures as Pierre-Simon Laplace and Joseph Fourier. Since their time, linearity has been tacitly assumed in much mathematical research, and an impressive edifice of linear theory has been elaborated. It is still an active area of research today.

The Historical Background—Nonlinear Systems

An important exception to the general rule of linearity was the set of differential equations derived for the motion of fluids. The prolific Swiss mathematician Leonard Euler published a paper in 1755 entitled "General Principles of the Motion of Fluids", in which he gave a set of partial differential equations to describe the motion of nonviscous fluids. These were subsequently improved by the French engineer Claude Navier, when in 1821 he published a paper giving equations which took viscosity into account. Independently, in 1845 the British mathematical physicist George Gabriel Stokes derived from somewhat different hypotheses the same equations in a paper entitled "On the Theories of the Internal Friction of Fluids in Motion" [Klin72]. These are the famous Navier-Stokes equations, a set of nonlinear partial differential equations which are quite general in their applicability to fluid motion (for example, to hydrodynamics and aeronautics) and are of the utmost importance in practical problems. In attempting to solve these equations, mathematical difficulties arise through the presence of terms containing products of velocities and their derivatives—the nonlinear terms.

In the 1920s an English meteorologist, Lewis Fry Richardson (who also influenced the history of fractals through his measurement of coastlines), made an attempt to calculate solutions of the equations by specifying initial and boundary conditions and using numerical methods. He actually visualised an orchestra of human computers carrying out the vast numbers of calculations under the baton of a mathematician. His real need was a powerful computer, though the advent of chaos puts up a warning sign to modellers who believe that knowledge of initial conditions and possession of differential equations automatically produces a legitimate result.

During the 1920s the appearance of electric devices such as amplifiers and oscillators further illustrated the need for nonlinear analysis. Following the Second World War, John von Neummann noted that the existing analytical methods of mathematics were unsuitable for nonlinear problems and was quick to see the potential offered by the emergent high speed computers. For von Neummann, they were heuristic devices which could be used to discern clues in the quest of constructing a properly based theory of nonlinearity [Perl77].

Nonlinear problems are notoriously difficult. In fluid dynamics, for example, the steady flow of fluids is fairly well understood. But turbulent motion, such as the fast movement of water over rapids, is a problem of outstanding complexity. Of particular interest is the onset of turbulence, the point at which steady flow gives way to turbulent flow. A present objective of the fluid dynamicist is the investigation of turbulence and its relation to chaos.

What is Chaos?

The principal feature of chaos is that simple deterministic systems can generate what appears to be random behaviour. Chaos can be observed in basic mechanical systems. One of these is a variant of the simple pendulum, in which the pivot point is moved up and down. For many pivot frequencies the pendulum swings back and forth as normal, but as the pivot is slowed down a point is reached where the motion of the pendulum becomes erratic and unpredictable. At this point the regular periodic behaviour of the pendulum gives way to chaotic motion. Nor is chaos restricted to physical examples, for it is present in quite elementary-looking mathematical equations. Perhaps the most well known case is the first order difference equation

$$x_{n+1} = \alpha x_n (1 - x_n)$$

used to model population growth [May76]. Here x_n is the 'population' at time n and is measured on a scale between 0 and 1. The presence of the x_n^2 term makes the equation nonlinear. Suppose an initial value x_0 is chosen. For values of the parameter α between 0 and 3, the resulting behaviour of x_n is stable—it approaches one fixed value as n tends to infinity (see Figure 1). But if α is given the value 3, instability begins. The first sign of this is the limiting behaviour of x_n oscillating between two values (see Figure 2). For further increases in α, the period 2 gives way to periods 4, 8, 16 and so on—the period doubling phenomenon. Finally, if α is increased to 4, the limiting values of x_0 appear to act chaotically. These examples are discussed by Mullin [Mull91].

The appearance of apparent randomness from determinism seems paradoxical at first, though some scientists believe that a deeper understanding of the chaos phenomenon has the potential of providing scientific explanation in cases where experimental results and present theories are at variance.

Sensitivity of Initial Conditions

An essential hallmark of chaos in nonlinear systems is the extreme sensitivity of the system to initial conditions. This means that two sets of conditions of a system which are initially very close together can give rise to widely different states in the long term. For an elementary physical example of this, consider the pinboard invented by Sir Francis Galton to illustrate probability (see Figure 3).

According to Newton's laws, balls dropped from the same position with the same initial velocity should follow the same path to the bottom. In practice it is unlikely that balls could be dropped from *exactly* the same position with *exactly* the same initial velocity. When the balls hit the pins the smallest differences are magnified, and balls take different paths.

Since in nature the initial conditions cannot be known exactly, but only with limited accuracy, it follows that the predictability of long term behaviour must fail for those nonlinear systems which exhibit chaos. The sensitivity of nonlinear systems to initial conditions is popularly know as the 'butterfly effect', because

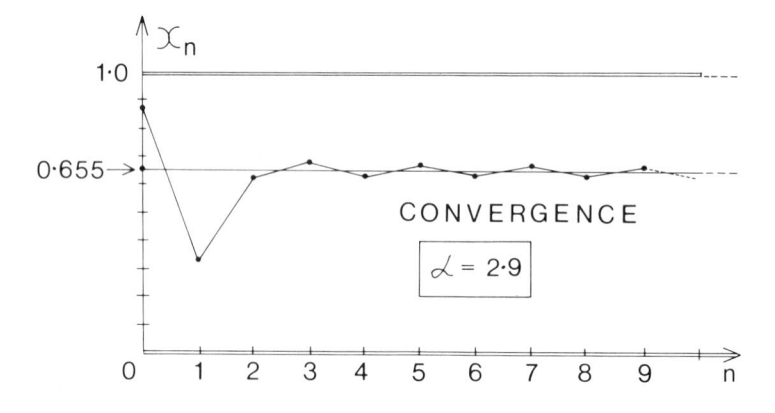

Figure 1. For $\alpha = 2.9$ and $x_0 = 0.875$, the value of x_n approaches the value 0.655 in a stable fashion.

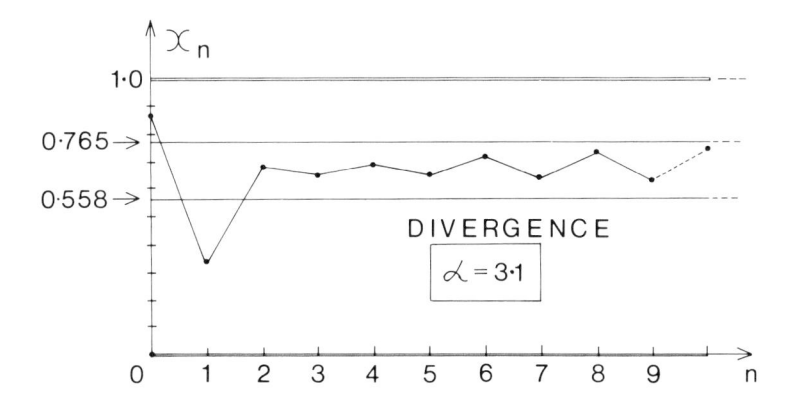

Figure 2. For $\alpha = 3.1$ and $x_0 = 0.875$, the value of x_n oscillates between values 0.558 and 0.765.

the single flap of a butterfly's wings would theoretically alter the initial conditions of a weather system and could thus give rise to drastically different weather patterns at a later time. Henri Poincaré, the eminent French scientist, was aware of this phenomenon when he wrote in 1903 regarding his study of planetary motion, "It may happen that small differences in the initial conditions produce very great ones in the final phenomena. A small error in the former will produce an enormous error in the latter. Prediction becomes impossible." [Crut86].

Certainly, this aspect of the chaos phenomenon is not necessarily bound up with random effects due to experimental error. It can be seen in simple mathematical processes such as the population model $x_{n+1} = \alpha x_n(1 - x_n)$, or in the

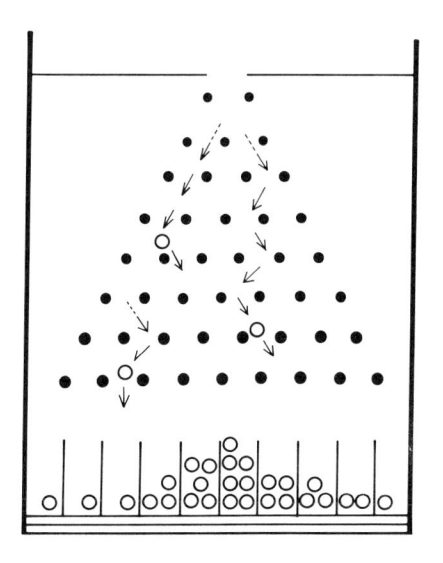

Figure 3. Galton's pinboard, which illustrates the sensitivity of the initial conditions to the paths taken by the balls.

following doubling transformation. In this example, we first write down a number on the continuous scale between 0 and 1 as a binary 'decimal'. We shall choose an irrational number a_0 and obtain an expression for it as an infinite sequence of binary 0's and 1's, such as

$$a_0 = .01011100111010111100\ldots$$

where this represents the number

$$a_0 = 0\left(\frac{1}{2}\right) + 1\left(\frac{1}{2^2}\right) + 0\left(\frac{1}{2^3}\right) + 1\left(\frac{1}{2^4}\right) + 1\left(\frac{1}{2^5}\right) + \cdots$$

Now consider the transformation which moves the 'decimal' point to the right and deletes the leading nonfractional part. Mathematically this is the function $x_{n+1} = f(x_n) = 2x_n \pmod{1}$. Apply the transformation to the number a_0 and keep applying it to obtain from

$$a_0 = .01011100111010111100\ldots$$

the successive numbers

$$a_1 = .1011100111010111100\ldots$$
$$a_2 = .011100111010111100\ldots$$
$$a_3 = .11100111010111100\ldots$$

and so on. In principle the system is fully deterministic, for we know the initial value a_0; and each subsequent value is absolutely determined by the function.

Now suppose we take a number b_0 which is close to a_0. A candidate for b_0 is

$$b_0 = .01011001010011010001\ldots$$

The numbers a_0 and b_0 agree in the first five places, but otherwise they are not connected with each other. If the function is applied to b_0 the results gradually diverge from the results obtained by applying the function to a_0, until the corresponding values lie on either side of the line through $x = 1/2$. The effect of $f(x_n) = 2x_n \pmod{1}$ on these values is to fold one value, and two divergent 'chaotic' time series are set in motion (see Figure 4). This phenomenon can happen no matter how close the initial points are together [Proc88b].

Trajectories and Phase Portraits

A pathway traced out by a solution of the population model equation is an example of a trajectory. Starting with a plot of the initial value, it shows the behaviour of the system over subsequent time.

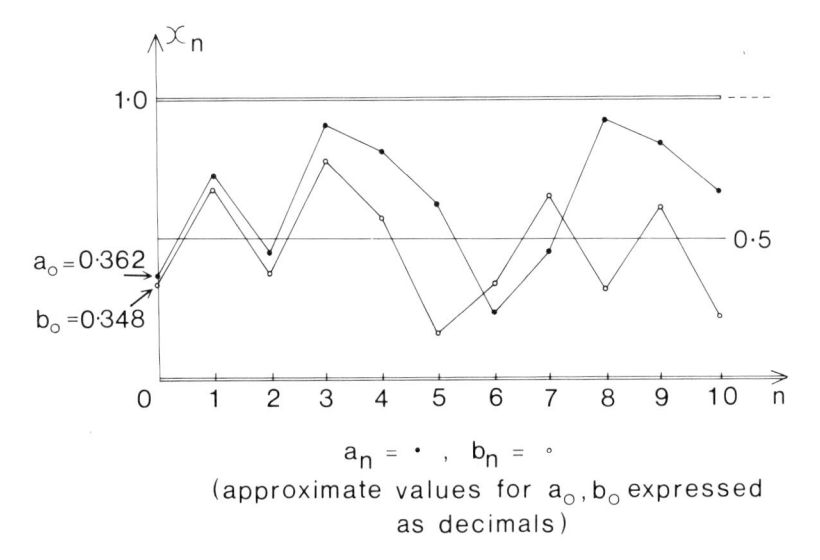

Figure 4. The two points a_0, b_0 are close together, but the trajectories followed by a_n, b_n diverge from each other.

The pictorial view of a dynamical system is usually described by the values of certain physical variables in a 'phase' space. For example, the motion of a simple pendulum can be described by specifying two variables at each instant of time: the angle made with the vertical, and the angular velocity of the bob. The main dynamical features of the system are known if the values of these variables are known for each instant of time. When these values are plotted in two-dimensional phase space starting with their initial values, a path is traced out called a trajectory. A collection of trajectories is called the phase portrait of the system.

Two elementary cases of phase portraits for the pendulum occur when the oscillatory motion is either undamped, or when it is damped due to energy loss caused by air resistance [Darb91]. The phase portrait of complete rotations of a pendulum is considered by Novak in a paper in this volume [Nova91].

The phase space of a simple pendulum with moving pivot is three-dimensional, since three variables in phase space are involved. In addition to the angle and angular velocity of the simple pendulum, another variable is the phase of the up and down motion of the pivot. The introduction of the pivot phase leads to the possibility of nonintersecting trajectories in the three-dimensional phase space, and it is shown that they are attracted to a chaotic attractor [Mull91].

A more complicated system involves two pendula coupled together. The bottom pendulum is free to swing about a central pivot in a plane at right angles to the top pendulum. The top pendulum is free to swing as a simple pendulum, but its pivot can also be driven up and down. In this case five variables (two angles, two angular velocities and the phase of the vertical driving mechanism) make the phase space five-dimensional [Pott91].

Nonlinear Ordinary Differential Equations

The differential equation is usually the most appropriate mathematical tool for analysing a dynamical system. In the case of the simple pendulum without damping, the differential equation is the classical

$$\ddot{x} + w^2 \sin x = 0$$

where x is the angle of swing and w is the frequency of swing. Although this is a nonlinear equation due to the presence of the $\sin x$ term, the motion of the pendulum is periodic, and the system does not give rise to chaotic motion. As has been seen, chaotic motion arises in the pendulum with the moving pivot; its underlying nonlinear differential equation is a modification of the equation of the damped pendulum

$$\ddot{x} + a\dot{x} + (1 + bw^2 \cos wt) \sin x = 0$$

The new term $a\dot{x}$ is related to the energy loss and the coefficient $(1 + bw^2 \cos wt)$ of $\sin x$ to the movement of the pivot.

Perhaps the most celebrated set of nonlinear ordinary differential equations is the Lorenz set. In the 1960's Edward Lorenz used a computer to model weather patterns, using a set of ordinary nonlinear differential equations

$$\dot{x} = a(y - x)$$
$$\dot{y} = bx - y - xz$$
$$\dot{z} = xy - cz$$

These are obtained as a reduction of the partial differential equations of convection; for this reason they are not universally accepted by fluid dynamicists as applicable to fluids in all circumstances. The variables x, y and z are related to 'convective overturning', horizontal temperature and vertical temperature; but the significance of the equations lies in their exhibition of chaos. It is a reminder of the subject's intrinsic mathematical difficulty that very little of the behaviour of the solutions can be proved with strict mathematical rigour [Spar82].

Strange Attractors

The classical dynamical systems can be thought of as possessing attractors in phase space. In the case of the damped pendulum, all trajectories are 'attracted' to the single point attractor at the origin (Figure 5). For the system described by Lorenz's equations an attractor exists, though it is not simply a point attractor. The way a trajectory winds around an attractor for this case is indicated by a two-dimensional projection (Figure 6). If trajectories of Lorenz's equations are plotted in three-dimensional phase space, Lorenz found that they are attracted

The Roots of Chaos—A Brief Guide   201

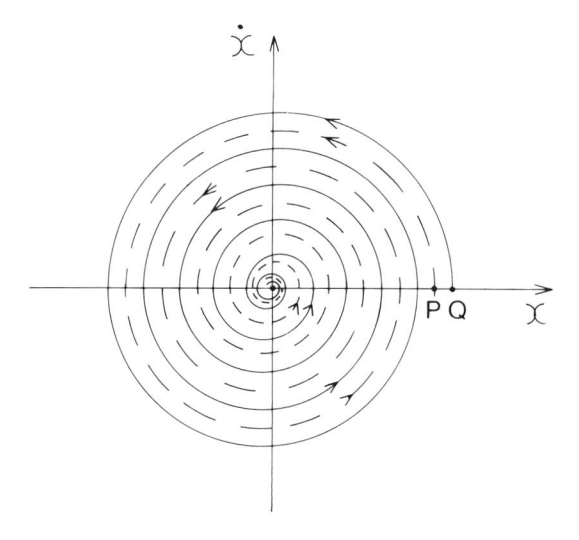

Figure 5. Two trajectories of the damped simple pendulum starting from initial conditions represented by points P and Q. Both trajectories are 'attracted' to the single point at the origin.

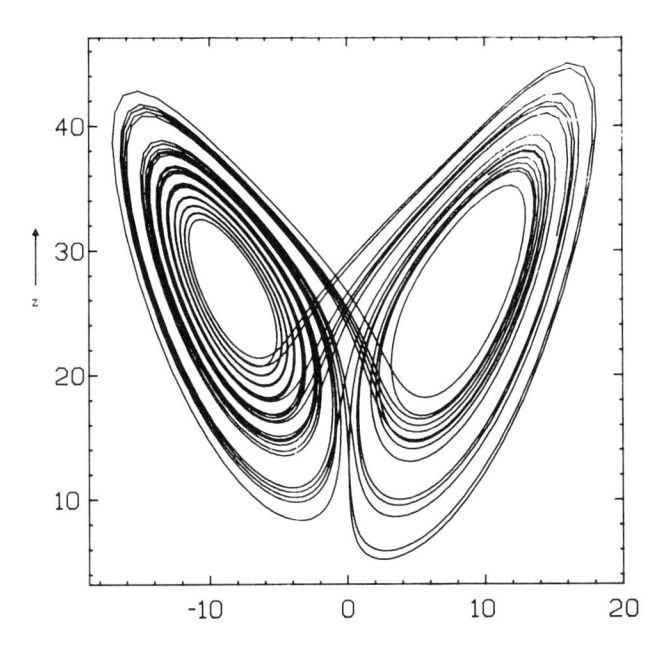

Figure 6. The Lorenz attractor for $a = 10$, $b = 28$, $c = 8/3$. The diagram shows a numerically computed solution to the Lorenz equations projected on the xz plane. (Reproduced from [Spar82].)

towards a bounded 'ellipsoid', which all trajectories eventually enter and from which they never emerge. A curious mathematical property of the ellipsoid is that it possesses zero volume. It is now known that the Lorenz attractor is an infinitely nested layered structure, and that the zero volume is consistent with cross sections through the layers being fractal in structure. Fractals do not have integer dimension like the more normal subsets of Euclidean space but have fractional dimension. Also, the more closely you look at fractals the more detail is observed, and they have the basic property of self-similarity [Jone91].

In contrast to the powerful computers of the 1980s and 1990s with their enhanced computer graphics, Lorenz made his discovery using a computer which could only make 17 calculations per second. In addition we have the delightful picture of 'string and sealing wax' research, in which the sensitivity of initial conditions' phenomenon was discovered by inadvertently using data with roundoff errors!

Empirical Science

Scientists have enthusiastically taken up the quest of finding strange attractors in the 'real' world. Experimental results also provide examples of the evolution of chaos which can be related to finite dimensional dynamical systems [Mull91].

Various experimental results have been reported in the scientific literature. In work on climatology using oxygen isotope records of deep sea cores spanning the past million years, the dimension of a climatic attractor has been reported to have a dimension of approximately 3.1 [Nico84]. Another team of researchers analyzed weather data over time scales ranging from 15 to 40 years and reported a dimension of the weather attractor between 6 and 7 [Tson88]. This result is far from being certain knowledge, and there is an ongoing debate on the validity of the techniques employed [Proc88a].

In each of these cases the dynamical behaviour of the system is described by a trajectory in a high dimensional phase space which settles on an attractor of comparatively low dimension. This is of scientific interest, since there is greater efficiency in studying dynamical systems if the number of variables can be reduced. An important instance of this occurs in the study of a flowing fluid, in which the phase space is of infinite dimension but in which, for certain simple physical set-ups, the motion may settle down on an attractor of low dimension. Viewed in this way, the topological description of the attractor holds the key to the dynamical behaviour of the system. One method of discovering the nature of a low dimensional system is to reconstruct an attractor using a time-delayed coordinate method. This technique allows a representation of the system to be constructed from a simple time series [Darb91].

In these developments computer graphics tools have had important influence. This has also been the case in other nonlinear theories of physics. One example is the remarkable solutions of the complicated set of nonlinear partial differential equations which govern magnetic monopoles where computer graphics has enabled their interaction to be modelled [Pott91].

In Conclusion

The biologist R.M. May has speculated on the reasons why chaos took so long to come to prominence. Given that it is so remarkable and derives from such elementary equations, why did it not blossom from the beginning of the century when Poincaré considered the effect, or in the 1960s when Lorenz's paper was first published? In retrospect, it is also surprising that the work of Stanislaw Ulam in the 1950s on iterating simple quadratic and other nonlinear transformations did not have a greater impact [Ulam76]. Apart from the need for powerful computers, May believes that chaos had to wait to be seen in systems which were simple enough for generalities to be perceived, and in contexts where practicalities could be appreciated [May87]. The emergence of chaos has much in common with other discoveries in mathematics which had to await the right conditions for their importance to be recognised. The case of matrix algebra is an obvious example of this historical evolution.

Chaos suggests new approaches to studying phenomena, but it also offers insights into the practice of modelling itself. Viewpoints are subtly altered, and in dynamics, for example, concepts such as continuity are replaced by fractals, while analysis is replaced by geometry, computers and number theory [Perc87]. These new approaches are also of practical significance, e.g., in the science of weather forecasting where chaos is used to assess the credibility of weather predictions. If small variation in the initial conditions on the model gives rise to very different predictions, then the forecast in this instance can be judged to be unreliable.

In studying chaos theoretically, mathematicians and physicists are having to look again at simple nonlinear systems. It may be conjectured that the level of generality and abstraction which has been gained in linear mathematics will take a long time to occur in nonlinear theories. The mathematician E.C. Zeeman has remarked on the "vast mountain of unsolved problems for the mathematicians to work on" [Zeem87a]. As the theory of nonlinearity unfolds, it is clear that discoveries will be made with the help of the computer, guided by a combination of intuition, analogy and accumulated mathematical and physical experience.

At present the subject is being pursued with fervour, and the application of chaos to many disparate areas is widespread, with much more prophesied. Some spectacular items include the chaotic tumbling of Hyperion, one of Saturn's more distant satellites whose orbit is smooth but which tumbles chaotically. The human heart beating regularly, followed by the onset of irregular beating and heart attack, is an area of medicine that fits into the schema of chaos research [Glas87]. The spread of infectious diseases is also another promising area of investigation. More controversial is the suggestion that chaos might be used to model schizophrenia [Zeem87b].

Chaos is genuinely interdisciplinary and brings together workers in scientific disciplines which were previously thought to have had little in common. This comes about by seeing chaos as a new way of analysing the results of experiments and is thus widely applicable to scientific areas which generate data. One interesting feature of the chaos advance is that pieces of abstract pure mathematics are now found to be relevant to further development.

Further Reading

The range of written material on chaos has grown extensively since the middle 1970s; it ranges from expository articles in such journals as *Nature* and *New Scientist*, to technical mathematics/physics journals such as *Nonlinearity* and *Physica D*. Though the ideas of chaos may be regarded as straightforward on one level, they quickly lead to a sophisticated and highly technical theory. The list of references given here is only intended as a foothold on the subject.

For the reader who wishes to gain an overview of chaos and experience some of the excitement as the new ideas have arisen, there is the popular book by James Gleick which is intended for the general reader [Glei88]. The papers presented to a Royal Society conference held in London in 1987 are very informative though written on a technical level [Berr87].

Articles in *New Scientist* are accessible and intended for the general reader and are written by experts in the field. More technical expositions can be found in *Nature*; the general reader would find much of interest especially in the early issues when authors concentrated on explaining the basic ideas. Articles in *Nature* can be used as a guide to the specialist scientific literature.

Acknowledgements. I have received helpful suggestions from among the contributors to this volume. I would also like to thank Professor A.D.D. Craik for his comments on the first draft of this brief guide. I am responsible for any errors, omissions and mistaken beliefs which remain.

REFERENCES

[Berr87]
  Berry, M.V., Percival, I.C., and Weiss, N.O., Eds., Dynamical chaos, *Proc. Roy. Soc. London*, Vol. 413, pp. 1–199, 1987.

[Crut86]
  Crutchfield, J.P., Farmer, J.D., Packard, N.H. and Shaw, R.S., Chaos, *Sci. Amer.*, Vol. 255, pp. 38–49, 1986.

[Darb91]
  Darbyshire, A.G., and Price, T.J., Phase portraits from chaotic time series, *this volume*, pp. 247—257, 1991.

[Glas87]
  Glass, L., Goldberger, A.L., Courtemanche, M., and Schreir, A., Nonlinear dynamics, chaos and complex cardiac arrhythmias, in [Berr87], pp. 9–26.

[Glei88]
  Gleick, J., *Chaos–Making a New Science*, London: Sphere, 1988.

[Jone91]
  Jones, H., Fractals before Mandelbrot, *this volume*, pp. 7—33, 1991.

[Klin72]
  Kline, M., *Mathematical Thought From Ancient to Modern Times*, New York: Oxford Univ. Press, 1972.

[Lans91]
Lansdown, J., Chaos, design and creativity, *this volume*, pp. 211—224, 1991.

[May76]
May, R.M., Simple mathematical models with very complicated dynamics, *Nature*, Vol. 261, pp. 459–467, 10 June 1976.

[May87]
May, R.M., Chaos and the dynamics of biological populations, in [Berr87], pp. 27–44.

[Mull91]
Mullin, T., Chaos in physical systems, *this volume*, pp. 237—245, 1991.

[Nico84]
Nicolis, C., and Nicolis, G., Is there a climatic attractor?, *Nature*, Vol. 311, pp. 529–532, 4 Oct. 1984.

[Nova91]
Novak, M.M., Relativistic particles in a magnetic field, *this volume*, pp. 225—236, 1991.

[Perc87]
Percival, I.C., Chaos in Hamiltonian systems, in [Berr87], pp. 131–143.

[Perl77]
Perlmutter, A., and Scott, L.F., Eds., *The Significance of Nonlinearity in the Natural Sciences*, New York, London: Plenum Press, 1977.

[Pott91]
Pottinger, D., Data visualization techniques for nonlinear systems, *this volume*, pp. 259—268, 1991.

[Proc88a]
Procaccia, I., Complex or just complicated, *Nature*, Vol. 333, pp. 498–499, 9 June 1988.

[Proc88b]
Procaccia, I., Universal properties of dynamically complex systems: the organization of chaos, *Nature*, Vol. 333, pp. 618–623, 16 June 1988.

[Spar82]
Sparrow, C., *The Lorenz Equations: Bifurcations, Chaos, and Strange Attractors*, New York, Berlin: Springer-Verlag, 1982.

[Tson88]
Tsonis, A.A., and Elsner, J.B., The weather attractor over very short timescales, *Nature*, Vol. 333, pp. 545–547, 9 June 1988.

[Ulam76]
Ulam, S.M., *Adventures of a Mathematician*, New York: Scribner, 1976.

[Zeem87a]
Zeeman, E.C., Chairman's Introduction (to Dynamical chaos), in [Berr87], pp. 3–4.

[Zeem87b]
Zeeman, E.C., General discussion (on Dynamical chaos), in [Berr87], p. 199.

Additional Readings

[Abra89]
Abraham, R.H., and Shaw, C.D., *Dynamics—The Geometry of Behaviour*, Vols. 1–4, Santa Cruz, CA: Aerial Press, 1989.

[Bai84]
Bai-Lin, H. (Ed.), *Chaos*, Singapore: World Scientific, 1984.

[Barn86]
Barnsley, M.F., and Demko, S.G., *Chaotic Dynamics and Fractals*, London: Academic Press, 1986.

[Barr88]
Barrow, J., Getting to grips with complexity, *New Scient.*, No. 1599, p. 63, 11 Feb. 1988.

[Belo89]
Beloshapkin, V.V., Chernikov, A.A., Natenzon, M.Ya., and Petrovichev, B.A., Chaotic streamlines in pre-turbulent states, *Nature*, Vol. 337, pp. 133–137, 12 Jan. 1989.

[Berr87]
Berry, M.V., Quantum physics on the edge of chaos, *New Scient.*, No. 1587, pp. 44–47, 19 Nov. 1987.

[Bohr87]
Bohr, T., and Cvitanovic, P., Chaos is good news for physics, *Nature*, Vol. 329, pp. 391–392, 1 Oct. 1987.

[Camp85]
Campbell, D., Crutchfield, J.P., Farmer, J.D., and Jen, E., Experimental mathematics: The role of computation in nonlinear science, *CACM*, Vol. 28, pp. 374–384, 1985.

[Carr83]
Carrigan Jr., R.A., and Trower, W.P., Magnetic monopoles, *Nature*, Vol. 305, pp. 673–678, 20 Oct. 1983.

[Chow88]
Chown, M., In the kingdom of chaos, *New Scient.*, No. 1639, pp. 56–57, 19 Nov. 1988.

[Cvit89]
Cvitanovic, P., *Universality in Chaos*, 2nd ed. Bristol, U.K.: Adam Hilger, 1989.

[Davi87a]
Davies, P., *The Cosmic Blueprint*, London: Heinemann, 1987.

[Davi87b]
Davies, P., The creative cosmos, *New Scient.*, No. 1591, pp. 41–44, 17 Dec. 1987.

[Deva87]
Devaney, R.L., *An Introduction to Chaotic Dynamical Systems*, Redwood City, CA: Addison Wesley, 1987.

[Ekel88]
Ekeland, I., *Mathematics and the Unexpected*, Chicago: Univ. of Chicago Press, 1988.

[Ford83]
Ford, J., Chaos at random, *Nature*, Vol. 305, p. 664, 20 Oct. 1983.

[Geis82]
Geisel, T., Chaos, randomness and dimension, *Nature*, Vol. 298, pp. 322–323, 22 July 1982.

[Guck82]
Guckenheimer, J., Noise in chaotic systems, *Nature*, Vol. 298, pp. 358–361, 22 July 1982.

[Guck83]
Guckenheimer, J., and Holmes, P., *Nonlinear Oscillations, Dynamical Systems, and Bifurcations of Vector Fields*, New York: Springer-Verlag, 1983.

[Harr86]
Harrison, R.G., and Biswas, D.J., Chaos in Light, *Nature*, Vol. 321, pp. 394–401, 22 May 1986.

[Hofs86]
Hofstadter, D.R., *Metamagical Themas: Questing for the Essence of Mind and Pattern*, Harmondsworth, U.K.: Penguin, 1986 (first published Basic Books, 1985).

[Hold83]
Holden, A.V., Chaos in complicated systems, *Nature*, Vol. 305, p. 183, 15 Sept. 1983.

[Hold85]
Holden, A.V., Frontiers of chaos, *Nature*, Vol. 316, p. 390, 1 Aug. 1985.

[Kloe76]
Kloeden, P., and Deakin, M., A precise definition of chaos, *Nature*, Vol. 264, p. 295, 18 Nov. 1976.

[MacD78]
MacDonald, N., Coupled oscillators in chaotic modes, *Nature*, Vol. 274, p. 847, 1978.

[MacD80]
MacDonald, N., Nonlinear dynamics, *Nature*, Vol. 283, pp. 431-432, 31 Jan. 1980.

[MacD82]
MacDonald, N., Noisy chaos, *Nature*, Vol. 286, pp. 843–844, 28 Aug. 1982.

[Madd83]
Maddox, J., Chaos theory infects civil engineers, *Nature*, Vol. 304, p. 115, 14 July 1983.

[Mand77]
Mandelbrot, B., *Fractals: Form, Chance and Dimension*, San Francisco: W.H. Freeman, 1977.

[Mand89]
Mandelbrot, B., Chaos, Bourbaki and Poincaré, *Math. Intelligencer*, Vol. 11, No. 3, pp. 10–12, 1989.

[Maso84]
Mason, J., Mathews, P., and Westcott, J.H., Eds., *Predictability in Science and Society*, Cambridge, U.K.: Cambridge Univ. Press, 1984.

[May89]
: May, R.M., The chaotic rhythms of life, *New Scient.*, No. 1691, pp. 37–41, 18 Nov. 1989.

[Moon87]
: Moon F., *Chaotic Vibrations: An introduction for Applied Scientists and Engineers*, New York: John Wiley and Sons, 1987.

[Mull89]
: Mullin, T., Turbulent times for fluids, *New Scient.*, No. 1690, pp. 52-55, 11 Nov. 1989.

[Mull89]
: Mullin, T., and Price, T.J., An experimental observation of chaos arising from the interaction of steady and time-dependent flows, *Nature*, Vol. 340, pp. 294–296, 27 July 1989.

[Murr89]
: Murray, C., Is the solar system stable?, *New Scient.*, No. 1692, pp. 60–63, 25 Nov. 1989.

[Nature82]
: Nature, What future for the chaos concept, *Nature*, Vol. 300, p. 311, 25 Nov. 1982.

[Nature83]
: Nature, Chaos, a problem for experiment, *Nature*, Vol. 303, p. 15, 5 May 1983.

[NewSci88]
: New Scientist, Boxed barium clarifes chaotic behaviour, *New Scient.*, No. 1628, p. 41, 3 Sept. 1988.

[Otti88]
: Ottino, J.M., Leong, C.W., Rising, H., and Swanson, P.D., Morphological structures produced by mixing in chaotic flows, *Nature*, Vol. 333, pp. 419–425, 2 June 1988.

[Palm89]
: Palmer, T., A weather eye on unpredictability, *New Scient.*, No. 1690, pp. 56–59, 11 Nov. 1989.

[Perc85]
: Percival, I.C., What is chaos, *Nature*, Vol. 317, pp. 629–680, 24 Oct. 1985.

[Peit86]
: Peitgen, H.O., and Richter, P.H., *The Beauty of Fractals*, Berlin: Springer-Verlag, 1986.

[Ruel80]
: Ruelle, D., Strange attractors, *Math. Intelligencer*, Vol. 2, pp. 126–137, 1980.

[Schu84]
: Schuster, H.G., *Deterministic Chaos*, Weinheim, Federal Republic of Germany: Springer-Verlag, 1984.

[Scot89]
: Scott, S., Clocks and chaos in chemistry, *New Scient.*, No. 1693, pp. 53–59, 2 Dec. 1989.

[Skel89]
: Skeldon, A., Mullin, T., and Pottinger, D.E.L., The five dimensional pendulum picture show, *New Scient.*, No. 1689, pp. 46–47, 4 Nov. 1989.

[Spie87]
Spiegel, E.A., Chaos: a mixed metaphor for turbulence, in [Berr87], pp. 87–95.

[Stew87a]
Stewart, I., The arithmetic of chaos, *Nature*, Vol. 329, pp. 670-671, 22 Oct. 1987.

[Stew87b]
Stewart, I., *The Problems of Mathematics*, Oxford, U.K.: Oxford Univ. Press, 1987.

[Stew89a]
Stewart, I., Portraits of chaos, *New Scient.*, No. 1689, pp. 42–47, 4 Nov. 1989.

[Stew89b]
Stewart, I., *Does God Play Dice?: The Mathematics of Chaos*, Oxford, U.K.: Blackwell, 1989.

[Sutt66]
Sutton, O.C., *Mathematics in Action*, London: Bell, 1966.

[Thom88]
Thompson, J.M.T., and Stewart, H.B., *Nonlinear Dynamics and Chaos*, New York: John Wiley and Sons, 1988.

[Viva89]
Vivaldi, F., An experiment with mathematics, *New Scient.*, No. 1688, pp. 46–49, 28 Oct. 1989.

[Wigg88]
Wiggins, S., Stirred but not mixed, *Nature*, Vol. 333, pp. 395–396, 2 Jun. 1988.

[Winf83]
Winfree, A.T., Sudden cardiac death–a problem in topology, *Sci. Amer.*, Vol.248, pp. 118–131, May 1983.

[Wolf83]
Wolf, A., Simplicity and universality in the transition to chaos, *Nature*, Vol. 305, pp. 182–183, 15 Sept. 1983.

[Wolf84]
Wolfram, S., Cellular automata as models of complexity, *Nature*, Vol. 311, pp. 419–424, 4 Oct. 1984.

Chaos, Design and Creativity

John Lansdown

Abstract

After briefly considering the role of randomness and unpredictability in science (which has a longer history than is sometimes thought), this paper goes on to look at randomness as a generative element in modelling artistic creativity. In particular, the visual aspects of chaotic functions and Iterated Function Systems (IFSs) are introduced. Consideration is given to the way in which IFSs might be used to facilitate a special method of designing—by so-called procedural modification. The paper concludes with an example of such use to create variations on one of the decorative designs by Scottish architect C.R. Mackintosh.

Introduction

Perhaps since the days of Maxwell (1831–1879)—or at least those of Poincaré (1854–1912)—it has been clear that apparently deterministic phenomena can sometimes display unpredictably random behaviour. The force of this knowledge led Maxwell to say that "The true logic of this world is the calculus of probabilities" and Poincaré to observe acidly that "Determinism is a fantasy due to Laplace". In his *Méthodes Nouvelles* of 1892, Poincaré showed conclusively that some regulated mechanical systems would develop chaotic behaviour in the course of time [Poin92]. However, it was not until E.N. Lorenz's now famous paper *Deterministic Nonperiodic Flow* [Lore63] that some of the fuller implications of this truth were realised.

The American Air Force funded Lorenz, a professor at Massachusetts Institute of Technology (MIT), to work on long-range weather forecasting techniques. In the course of this work he devised a completely deterministic system to represent thermal convection in simplified form. The system consisted of the three ordinary differential equations

$$\frac{dX}{dt} = -s(X - Y)$$

$$\frac{dY}{dt} = -XZ + rX - Y$$

$$\frac{dZ}{dt} = XY - bZ$$

On the face of it these equations should model the behaviour of thermal convection in a straightforward and predictable way. In addition, slight changes of the input parameters should produce slight changes in the output values. But Lorenz found that, for some values of the parameters s, r and b and for some initial conditions, these equations exhibit irregular and unpredictable, yet bounded, behaviour and further, that tiny differences in the starting values of the parameters would produce wildly different output (Figure 1). This discovery has had profound implications for the development of weather forecasting models which—to be useful over a reasonable period of time—turn out to need infinitely accurate precision in their starting values. Pippard [Pipp85] summarises the difficulty thus:

It has been the habitual claim of physicists that they could make predictions whose verification underpinned the laws and conferred on science a validity that no other branch of learning can aspire to. Was this a delusion? Of course not, but the claim may have been over optimistically expressed ... There is little hope of ever knowing whether this day fortnight will be wet or dry—only that some weather patterns are more likely than others.

Random and Pseudorandom Numbers

It should come as no surprise to anyone who knows about computing that completely deterministic equations can sometimes produce apparently random behaviour. It is by use of exactly this principle that pseudorandom numbers are generated. The literature abounds with simple equations designed to create streams of statistically random numbers with minimal computation. Virtually all these are of the basic form

$$f(X(n+1)) = g(X)f(X(n)) \bmod p \qquad p, \text{prime}$$

That is, they are bounded, recursive and each successive number is dependent on the value of the one that went before. Indeed, what seems to be becoming

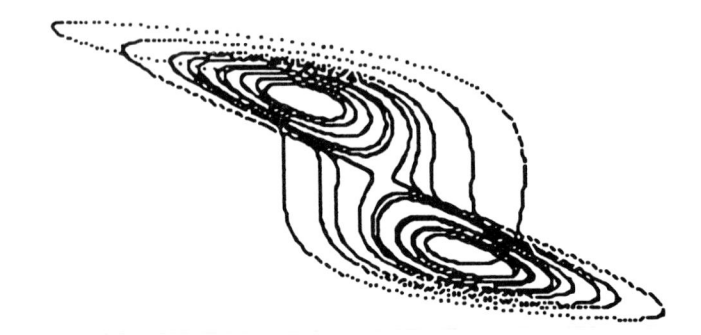

Figure 1. The Lorenz attractor.

clear is that, in any bounded nonlinear system where the state of the system at time $t+1$ is dependent on its state at time t (that is, most systems), one of the modes of behaviour is likely to be chaotic in the sense that its current state is unpredictably related to its previous states.

Incidentally, as *any* iterative equation can be converted to a recursive one, it is likely that there are infinitely more 'chaotic' equations than any of us have presently examined. It would be interesting to know how to find families of these to fit certain prerequirements. For instance, a recently published equation by Stewart [Stew89] opens up a fascinating development in symmetrical imagery based on the equilateral triangle (Figure 2).

The Visual Effect of Some Recursive Equations

All this would be of interest only to mathematicians, scientists and weather forecasters were it not for one thing. This is that, for some values of input parameters, interesting (and sometimes beautiful) patterns can arise when we freeze the time course of development of these sorts of equations. Thus, even though they produce streams of more or less random numbers when considered number by number, our perceptual mechanisms are able to give them some order when we view them as plotted points. This makes the subject of growing value to artists and designers. For example, the equations Martin [Mart89] investigates, namely

$$X(n+1) = Y(n) - \text{sgn}(X(n))\sqrt{|bX(n) - c|}$$
$$Y(n+1) = a - X(n)$$

for the constants $a = 3.5$, $b = 1.2$ and $c = 0.0$ produces successive values of Y, as shown in Figure 3a, and successive values of X, as shown in Figure 3b. These values would probably not be useful as pseudorandom numbers because they illustrate some form of order (for instance, that the numbers tend to lie in roughly regular formations on either side of a middle line). However, no individual number is repeated, and successive numbers are random within this overall regularity.

Figure 2. Stewart attractor based on an equilateral triangle.

214   John Lansdown

(a)   (b)

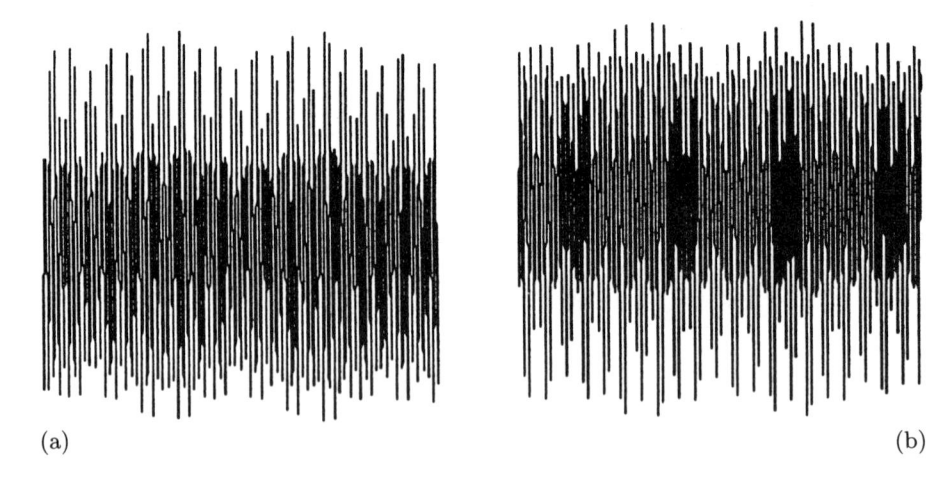

Figure 3. (a) shows some of the y values and (b) shows some of the x values used in generating Figure 4. They resemble the appearance of white noise plots.

When the successive pairs of X and Y are plotted in the XY plane, an unexpectedly beautiful pattern emerges, as Figure 4, the pattern after about 5,000 iterations, illustrates. If we look at an area of this plot in detail we see the essential irregularity of the points; Figure 5 shows the plot after 30,500 iterations. In addition—something that cannot be illustrated in a still picture—the points do not appear in a regular sequence as the drawing is being built up. The points are plotted in apparently random order, sometimes here, sometimes there in a highly unpredictable way, and it is only after a little while that we recognise that any pattern is emerging. Furthermore, the whole takes on different but related patterns as the number of iterations increases—at 500 iterations it is one pattern, at 5,000 another, at 50,000 yet another, and so on (as far as we know) ad infinitum.

For those who are new to such calculations and who want to explore the patterns on their own microcomputers, do not think that it is necessary to set up huge arrays to accommodate the successive values of X and Y as is sometimes implied in the literature—for instance, in the recent otherwise useful and practical book by Reitman [Reit89]. Because each pair of values can be calculated from the pair immediately before them, it is only necessary to store $X(i), Y(i)$ and $X(i+1), Y(i+1)$ in each iteration. After each point is plotted, the values must then, of course, be swapped so that the previous $X(i+1), Y(i+1)$ in one iteration become $X(i), Y(i)$ in the next, and so on. In this way arrays only of dimension 2 are needed—if indeed they are required at all. Thus, without fear of storage overflow any number of iterations can be generated up to the limit of one's patience. Note, though, that the size of images might vary greatly and unexpectedly with slightly different values of the parameters. If the values are not stored to enable the maximum and minimum values to be ascertained before plotting, it will be necessary to do a few short experimental runs to establish the drawing limits.

Chaos, Design and Creativity   215

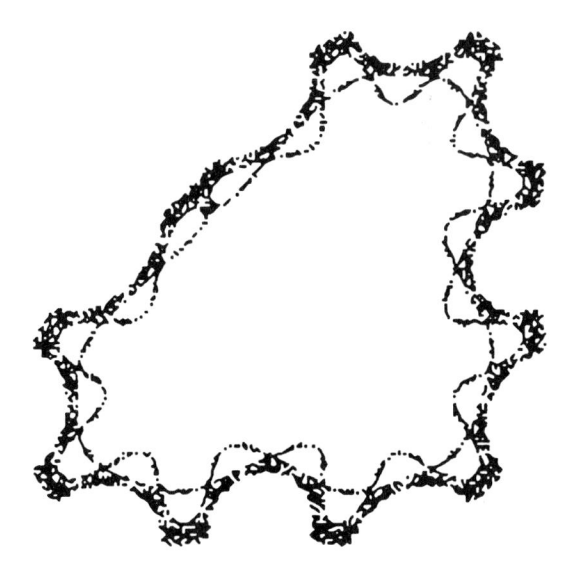

Figure 4. Martin's equation after 5,000 iterations.

The range and diversity of patterns that can arise just from varying the parameters in Barry Martin's equations alone [Mart89] make such recursive functions of great visual attraction to artists and designers. As it happens, this attractiveness (no pun intended) was recognised some time ago and predates the current widespread scientific interest in chaos theory. Most of the well-known literature on chaos theory dates from the early- and mid-1980s (see, for example, [Cvit89]), but at the Stockholm IFIP Information Processing 74 Conference, Gumowski and Mira [Gumo74] showed some fascinating images of recurrence equations in a session which I chaired (see Figure 6).

In September of that year I wrote up the talk that Gumowski gave [Lans74]. I did not, of course, mention 'chaos', because the term was not used in this context

Figure 5. Detail of Figure 4 after about 30,500 iterations.

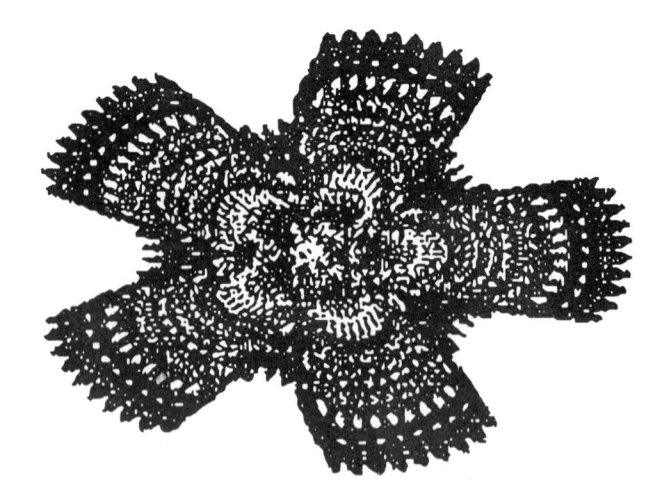

Figure 6. An example of one of Gumowski and Mira's recurrence equation images.

until Li and Yorke's paper of 1975 [Li75]. Later, in a popular personal computer magazine I published a short BASIC program to generate copies of Gumowski and Mira's images [Lans78] and commented on the apparently random order of generation, as well as the unpredictability and wide variability of the effects of parametric changes. This method was used by myself and a number of other computer artists to make pictures in the 1970s.

In passing, it is interesting to speculate whether Li and Yorke were doing us all a service in choosing the term 'chaos' to describe the effects resulting from this sort of work. I tend to think not. It is clear that what most of us are looking at here is not true chaotic behaviour but simply behaviour of higher than normal orders of complexity—even many of the most chaotic pictures seem to have some underlying visual structure. The fact that we can make sense of it at all is due to the remarkable pattern-resolving features of our Gestalt perceptual mechanisms.

Implications for Art and Design

To understand what makes all this of interest to those of us who use computers to study and simulate human artistic and design creativity, it is necessary to look at some of the past developments in this field. Over the past thirty or so years of its existence, visual computer art has progressed along three divergent paths:

> The first (and, in fact, the most recent) lies in the pursuit, via paint systems, of tools which enable artists to draw and paint their images *manually* but with computer assistance.
>
> The second, related to the first, lies in devising *programmed* techniques for drawing and painting images.
>
> The third lies in developing techniques for the *automatic creation* of images.

I do not intend to say more about the first two of these methods. They are both forms of the same thing: computer-assisted art and design. In using either of these first two methods, artists work toward concrete realisation of their mental images. I do not mean to suggest by this that these mental images are fully formed at the outset and are then simply transferred to the computer screen. On the contrary, as with conventional art they develop—and sometimes radically change—in the process of interaction with the tools and media [Lans89b]. But the approach manifested in the first and second techniques is quite different from that of the third. In the third approach, one starts not with an idea of an image to be realised but with a generating method to be explored. Thus, artists using the third approach are as much interested in the process of image making as in its outcome. They wish to see the effect of applying certain 'creative' techniques. Sometimes they generate large amounts of output and, using the same sort of artistic judgement they would apply to work created in a 'normal' way, select one or more items to show. In other cases, it is sufficient that the generative process is properly illustrated.

Elsewhere [Lans70; Lans89a] I have outlined a linguistic model for this sort of approach. As Figure 7 shows, I assume that we can divide the model's elements into four parts:

the vocabulary;
the grammar;
the presentation;
the selector.

This is by no means the only model that could be devised. Because of its comparatively limited applicability, it is unlikely to be the best. It has the merit here, however, of allowing us to discuss the parts separately. The *vocabulary* comprises the set of permissible individual words that can be used. 'Words' here can mean anything from the real words of a natural language for, say, poetry

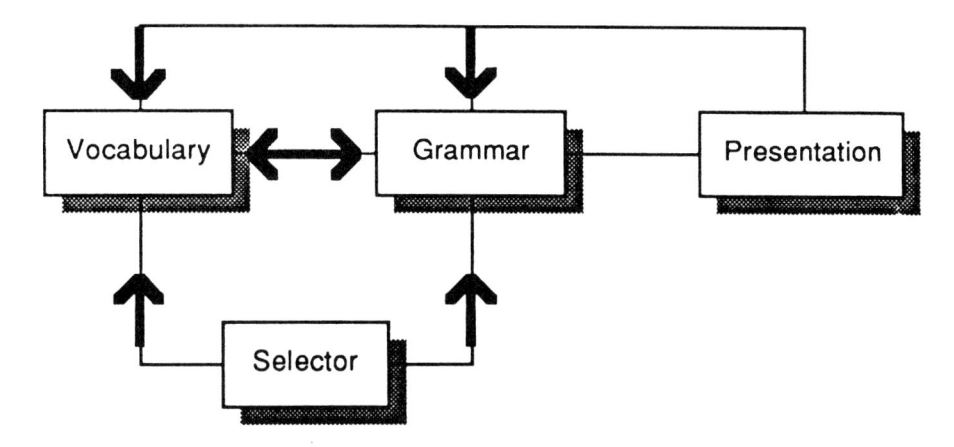

Figure 7. A linguistic model of creativity.

generating programs, to points, lines or planes in space for programs to generate pictures. The *grammar* is the set of rules that allows certain word combinations and disallows others. The *presentation* is the medium through which the output is to be manifested. Clearly, different media require different vocabularies, grammars and, possibly, selectors too. The *selector* is the mechanism for determining which elements of the vocabulary are to be used at any particular time.

The Role of Randomness

In natural language the selector is usually *meaning*. That is to say, I choose which words I want to use in accordance with the meaning I wish to convey. I use my imagination and intuition to make the choice of words as interesting and meaningful as I can, within the restrictions of the grammar. In visual communication similar conditions apply. But computers are not in a position to determine meaning themselves and have no imaginations or intuitions, so, in using them, some different mechanism for choosing appropriate elements of vocabulary has to be employed. Historically, this mechanism has been a random number generator. Thus, in creating an art work devised by an automatic generating method, particular elements of vocabulary are chosen randomly from a set of elements permitted by the grammar and presentation. Note that, unlike in day to day speech, there is no need for the grammar to insist that the selection of words is done in the order in which they are to be presented. In computer generated works—even those presented in real time—it is perfectly feasible, and in many cases highly desirable, for the order of selection to be very different from the order of presentation. For example, in musical composition or dance choreography, where a particular ending may be desired, it is very convenient to generate the work from end to beginning, even though it will be presented the right way around.

The reason that a random number generator is frequently used as the selector is that it can sometimes be constrained to make the unexpected choices and juxtapositions which simulate those made by imaginative people. Of course, this cannot happen if just zeroth-order randomness is employed. By zeroth-order randomness I mean randomness as manifested directly by the output of a random number generator. If we simply use a stream of random numbers without the mediation of a suitable, and probably complex, grammar, we have little chance of creating any meaning. This is because random number generators are designed to output streams of numbers in which each one is effectively independent of its predecessor. Any given number is used to generate its successor, but individual numbers apparently have no 'memory' of this fact. However, it is a characteristic of meaning that consecutive elements are intimately bound to one another. Each word in a sentence and each sentence in a paragraph, as it were, remembers the meaning that its predecessors convey.

Thus, in order to generate images having meaning, we have to combine use of the grammar and selector to simulate memory of what has gone before. This gives essentially an *n*th-order randomness. Various methods are available for achieving this effect. Examples are production rule based techniques [Stin78; Stin79; Smit84; Prus86]; life and death techniques [Gard71]; cell address ideas [Saun72; Dewd86]; cellular automata [Wolf84; Toff87]; and so on—see the papers by Lansdown [Lans88; Lans89a] for descriptions of some of these. Importantly, chaotic equations of the type that we have discussed above also have this characteristic. They seem to embody both a grammar and a selector. They have, effectively, an infinite memory of what has gone before. Sometimes this combination can produce striking and surprising results

Fractals and Iterated Function Systems

Fractals too have a particular coherence brought about by the special interrelationships of their parts—interrelationships which I have characterised in the last section as a form of memory. The use of fractals to create graphic images of such things as the Julia and Mandelbrot sets [Peit86; Dewd87] as well as of landscapes is too well known to need further rehearsal here [Mand82]. Fractals can be used to create images of great beauty and are a constant source of inspiration to some computer artists. Iterated Function Systems (IFSs) [Barn88] on the other hand have not been so widely explored, and their potential as devices for creating art and design images is only just starting to be understood. I have been looking lately at the use of Barnsley's Collage theorem and IFS to see how they might be used as an element in the procedural prototype design method which I have described elsewhere [Lans87].

A rough description of the Collage technique is that a complete image is built up of constituent parts which are transformed copies of one another. The transformations are governed by the application of sets of six parameters, which take care of translation, shearing, rotation and scaling. To present the image as an IFS, a further parameter is also needed for each transformation. This is a probability factor, which determines the density distribution of the random points that make up each part of the image. This parameter can be calculated more or less automatically from the others and is related to the areas of each part. The craft of the technique lies in discovering those parts of the image which have the greatest potential for re-use under the transformations. This is, of course, comparatively easy to do if the parts of the object to be depicted have an obvious degree of self-similarity—like the parts of a leaf or a tree. Harder problems exist if they lack self-similarity. However, it is a consequence of Barnsley's theorem that the transformable parts exist for all images.

The fact that powerful parameters come into play makes this technique, on the face of it, a suitable candidate for *procedural variation*. That is, given that you have sets of parameters which describe a particular image, what happens when you vary the parameters either in a random or in a grammar-controlled way? Note that for a 'sensible' well-behaved variation the parameters are not all independent of one another. The parameters a, b, c, d in Eq. (1) are interrelated in the manner of Eq. (2)

$$f(X) = aX + bY + e$$
$$f(Y) = cX + dY + f \quad (1)$$
$$a = S_h \cos(\theta)$$
$$b = -S_v \sin(\phi)$$
$$c = S_h \sin(\theta)$$
$$d = S_v \cos(\phi) \quad (2)$$

They take care of scaling, shearing and rotation and are thus governed by the horizontal and vertical scaling factors and the angles of rotation. So, strictly, one of a, b, c and d should not be varied independently of its partners. However, as artists and designers we are not interested in just the 'sensible' variations. We also want to see what happens when we do odd things: when we permute the values, say, or make some of them negative, or give one a new value.

By means of simple modifications to the published IFS and Collage programs, you can easily perform such procedural modifications. Figure 8, for example, shows random variations of the parameters of Barnsley's standard fern. Here, no attempt has been made to modify the probabilities based on the areas of the transformed shapes. Simply one, two or three parameters have been changed more or less arbitrarily.

A Design Variation

More serious investigations can be made using these techniques. I have examined concepts using, among other things, the designs for the finials on top of Charles Rennie Mackintosh's Glasgow School of Art railings as prototypes [Buch89]. In

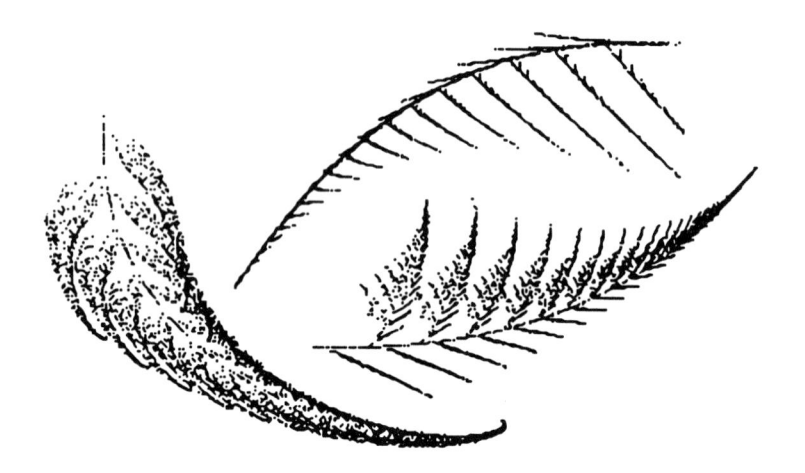

Figure 8. Random parametric variations of a fern IFS.

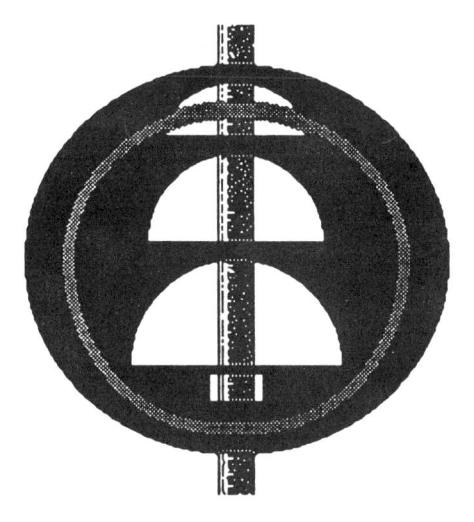

Figure 9. Simplified prototype of Mackintosh's railing finial.

common with many of the designs of the Arts and Craft Movement at the beginning of this century, these artefacts display something of the fascinating self-similarity of plants and other natural objects, so it is not surprising that they could be candidates for IFS. Taking one of these finials (Figure 9), we can describe it in terms of an IFS and then vary the parameters. These give rise to variations like those of Figure 10. The pictures you see there are not, of course, direct output from the IFS programs. They are drawn using the IFS output as a guide.

Concluding Remarks

From all these and other experiments, it is clear that chaos and fractal theory, particularly as manifested in the Collage Theorem, are of great potential significance to the artist and designer. By somehow embodying the grammar and

Figure 10. Some procedural variations on Figure 9 suggested by IFS output.

selector parts of the linguistic model, they allow us to create images at least comparable with that mixture of unexpectedness, order, complexity and irregularity which characterises more conventional art and design imagery.

Further, by revealing to us the fineness of line between order and chaos, they give us clues for devising new models of artificial creativity and, perhaps, greater understanding of human creativity as well. I look forward to seeing these developments.

REFERENCES

[Barn88]
Barnsley, M.F., *Fractals Everywhere*, San Diego: Academic Press, 1988.

[Buch89]
Buchanan, W., Ed., *Mackintosh's Masterwork: The Glasgow School of Art*, Glasgow: Richard Drew, 1989.

[Cvit89]
Cvitanovic, P., Ed., *Universality in Chaos* (2nd ed.), Bristol, UK: Adam Hilger, 1989.

[Dewd86]
Dewdney, A.K., Wallpaper for the mind: computer images that are almost, but not quite, repetitive, *Sci. Amer.*, Vol. 255, No. 3, pp. 14–23, 1986.

[Dewd87]
Dewdney, A.K., Beauty and profundity: The Mandelbrot set and a flock of its cousins called Julia, *Sci. Amer.*, Vol. 257, No. 5, pp. 140–145, 1987.

[Gard71]
Gardner, M., On cellular automata, self-reproduction, the garden of Eden and the game of life, *Sci. Amer.*, Vol. 224, No. 2, pp. 112–117, 1971.

[Gumo74]
Gumowski, I., and Mira, C., Point sequences generated by two-dimensional recurrences, in *Information Processing 74*, Amsterdam: North-Holland, pp. 851–855, 1974.

[Lans70]
Lansdown, J., Computer art for theatrical performance, *Proc. ACM Internat. Comput. Sympos.*, Bonn: ACM European chapters, pp. 717–735, 1970.

[Lans74]
Lansdown, J., Computer art is alive and well, *Comput. Bulletin*, Vol. 2, No. 1, pp. 10–11, 1974.

[Lans78]
Lansdown, J., Gumowski, *Creative Computing*, pp. 88–89, June 1978.

[Lans87]
Lansdown, J., The creative aspects of CAD: A possible approach, *Design Studies*, Vol. 8, No. 2, pp. 76–81, 1987.

[Lans88]
Lansdown, J., Computer Graphics: A Tool for the Artist, Designer and Amateur, in *Advances in Computer Graphics III*, de Ruiter, M.M., Ed., Berlin: Springer-Verlag, pp. 147–175, 1988.

[Lans89a]
Lansdown, J., Generative Techniques in Graphical Computer Art: Some Possibilities and Practices, in *Computers in Art, Design and Animation*, Lansdown, J., and Earnshaw, R.A., Eds., New York: Springer-Verlag, pp. 56–79, 1989.

[Lans89b]
Lansdown, J., Using the computer to augment design creativity, CASCAAD Working Paper, Middlesex Polytechnic, Barnet, UK, 1989.

[Li75]
Li, T.-Y., and Yorke, J.A., Period three implies chaos, *Amer. Math. Monthly*, Vol. 10, No. 82, pp. 985–992, 1975.

[Lore63]
Lorenz, E.N., Deterministic nonperiodic flow, *Jour. Atmospheric Sci.*, Vol. 2, No. 20, pp. 130–141, 1963.

[Mand82]
Mandelbrot, B.B., *The Fractal Geometry of Nature*, New York: W.H. Freeman, 1982.

[Mart89]
Martin, B., Graphic Potential of Recursive Functions, in *Computers in Art, Design and Animation*, Lansdown, J., and Earnshaw, R.A., Eds., New York: Springer-Verlag, pp. 109–129, 1989.

[Peit86]
Peitgen, H.-O., and Richter, P.H., *The Beauty of Fractals*, Berlin: Springer-Verlag, 1986.

[Pipp85]
Pippard, A.B., *Response and Stability*, Cambridge, UK: Cambridge University Press, 1985.

[Poin92]
Poincaré, J.H., *Les Méthodes Nouvelles de la Méchanique Celeste*, Paris: Gauthier-Villars, 1892.

[Prus86]
Prusinkiewicz, P., Graphical applications of L-systems, in *Graphics Interface 86*, Goldberg, M., and Woodham, R.J., Eds., Toronto: Canadian Information Processing Society, pp. 247–253, 1986.

[Reit89]
Reitman, E., *Exploring the Geometry of Nature*, Summit, PA: Windcrest, Blue Ridge, 1989.

[Saun72]
Saunders, R., A description and analysis of character maps, *Comput. Jour.*, Vol. 15, No. 3, pp. 160–169, 1972.

[Smit84]
Smith, A.R., Plants, fractals and formal languages, *Comput. Graph.*, Vol. 18, pp. 1–10, 1984.

[Stew89]
Stewart, I., Portraits of chaos, *New Scient.*, Vol. 124, No. 1689, pp. 42–47, 1989.

[Stin78]
Stiny, G., and Mitchell, W.J., The Palladian grammar, *Environ. and Planning B*, Vol. 5, pp. 5–18, 1978.

[Stin79]
Stiny, G., A generative approach to composition and style in architecture, *Proc. Parc79 Conf.*, Berlin: AMK, pp. 435–446, 1979.

[Toff87]
Toffoli, T., and Margolus, N., *Cellular Automata Machines*, Cambridge, MA: MIT Press, 1987.

[Wolf84]
Wolfram, S., Cellular automata as models of complexity, *Nature*, Vol. 311, No. 4, pp. 419–424, 1984.

Relativistic Particles in a Magnetic Field

Miroslav M. Novak

Abstract

A charged particle in the field of a plane wave travelling at right angles to the static magnetic field exhibits complex dynamics. Competition between the rotational and translational motions leads to the presence of stochastic channels, which separate regions where the motion is stable. Visualization techniques are used to improve our understanding of the inherent complexity.

Introduction

The purpose of this study is to investigate the behaviour of charged particles in the presence of an electromagnetic wave propagating in a static magnetic field, and to visualize the particles' energy dependence in different parameter regimes. The underlying process belongs to the group of Hamiltonian systems, so to begin with, let us briefly review some of the notions pertaining to this category.

Observation of the time history of the movement of the particle in a field can lead to an irregular motion that shows no clear pattern. Based on this observation alone, one cannot ascertain that the motion is chaotic as, if observed for a sufficiently long time, the trace may display long term regularity. Furthermore, the presence of quasi-periodic oscillations (when the ratio of frequencies is an irrational number) can result in an apparently chaotic motion. In order to reduce the possibility of incorrect identification, it is frequently useful to adopt some results from analytical mechanics.

In general, a system is characterized by generalized coordinates q_i and generalized momenta p_i, where the subscript denotes individual particles. The instantaneous state of a system can then be represented by a point (q_i, p_i) in phase space. Assuming that there are N particles in the system, the dimensionality of phase space is $6N$. As the state of the system changes, the point representing it moves in phase space. In time, this point follows a path which is known as a trajectory. From the shape of the trajectory one can infer the overall behaviour of the system. For instance, if the shape of the trajectory is a closed curve, the motion is periodic. This result also applies to a seemingly intersecting but closed trajectory, which describes the motion with a subharmonic period. On the other hand, open-ended trajectories, whereby the particle always follows a different path, indicate the possibility of deterministic but unpredictable behaviour [Moon87]. Let us look at this concept in more detail. The form of the trajectory can be found from the canonical equations of motion

$$\frac{\partial q_i}{\partial t} = \frac{\partial H(q_i, p_i)}{\partial p_i}$$
$$\frac{\partial p_i}{\partial t} = \frac{-\partial H(q_i, p_i)}{\partial q_i} \quad (1)$$

where ∂ stands for the partial derivative, H represents the Hamiltonian of the system, and $i = 1, 2, \cdots, 3N$.

A small volume of phase space contains a number of points, each of which represents a different but similar state of the system. As the system evolves, this phase volume moves around phase space, and in the absence of dissipation it retains its volume. This implies that two points initially close together will keep this proximity at a later stage. On the other hand, if the shape of the volume changes and becomes progressively more distorted, with time it completely disperses. Consequently, two initially close points are no longer together, the initial conditions lose their role and we are faced with an irregularity.

The trajectories that display the movement of the point continually do not supply sufficient information to establish the presence of chaos. Much more informative is the technique of Poincaré mapping, whereby the continuously updated trajectory is replaced by a suitably chosen sampling of the motion in phase space. In a harmonically driven system, a convenient sampling time interval is proportional to the perturbation frequency. The resultant map consists of a cluster of points that can now be analyzed. Several possibilities can arise:

(a) if the map consists of a finite number of discrete points, the particle executes harmonic oscillation;

(b) if a closed curve is formed, the motion is quasi-periodic, consisting of several oscillations with incommensurate frequencies;

(c) random-like distribution of points implies a probable chaotic motion.

One of the very important contributions in studying the stability of motion in general is due to Kolmogorov, Arnol'd and Moser and is often referred to as the KAM theorem [Arno63; Walk69]. The theory deals with nonlinear Hamiltonian systems. These are generally nondissipative problems, in which the N degree-of-freedom system is described by $2N$ first-order differential equations of motion. The KAM results describe the existence of stability regions that exhibit regular motion. The theory predicts that small perturbations will not completely suppress the stable motion. Suppose that a system is described by a Hamiltonian with N invariants. The corresponding trajectory of the system describes an N-dimensional torus. The presence of a perturbation V is absorbed by the Hamiltonian, which can now be written as

$$H = H_0 + V(t) \quad (2)$$

where H_0 is the Hamiltonian in the absence of any perturbation. Classically, the unperturbed Hamiltonian is simple, independent on time, while the perturbation is small but can vary with time. Assuming that certain conditions are fulfilled, the KAM theory indicates that the small perturbation will not destroy the trajectories on the torus, although they may become distorted. A further implication is that phase space in Hamiltonian systems always contains regions of stability, which may be, nevertheless, very small.

Consider now a simple mathematical pendulum, represented by an equation

$$\frac{d^2 x}{dt^2} + w^2 \sin x = 0 \tag{3}$$

where w is the angular frequency. Plotting during a time interval the angle x and the angular velocity dx/dt for differing initial conditions leads to a set of trajectories, depicted in Figure 1. As shown, the phase portrait describes two distinct modes of behaviour available to the pendulum, namely oscillation and rotation. The central part, containing the closed curves, represents the oscillations; the momentum as well as the deflections vary between the fixed boundaries. As the pendulum passes its equilibrium position ($x = 0$), its velocity is the largest; the maximum deflection occurs when the velocity vanishes. As the initial momentum of the pendulum is increased, the area enclosed by the trajectory also increases, until the limiting closed curve, the so called separatrix, is reached. Further increase in the momentum forces the pendulum into rotation, a fact represented by the open curves in Figure 1. The boundary between the two modes of motion, oscillation and rotation, is the separatrix; it passes through the unstable points $-\pi$ and π.

The presence of a periodic perturbation modifies Eq. (3) as

$$\frac{d^2 x}{dt^2} + w^2 \sin x = \alpha \sin(kx - \Omega t) \tag{4}$$

where α denotes the perturbation strength, and k and Ω stand for the wave vector and frequency of periodic perturbation. Small values of the perturbation parameter destroy the trajectories in the vicinity of the separatrix, and chaotic regions appear, as seen in Figure 2. The existence of such stochastic bands is common to Hamiltonian systems; they divide phase space into regions of different properties. Further increase in the perturbation parameter α destroys the original separatrix completely, and a new arrangement in phase space appears.

In the next section we find a basic description of a charged particle in an electromagnetic field. The section introduces the forces acting on the particle and leads to the associated differential equations describing its motion. The section on determination of the trajectory presents the method of analysis and considers several issues for obtaining successful representations in phase space. It presents some of the results that are obtained with different starting values and illustrates the effect on the trajectory using different parameters. Initial conditions play a vital role in the behaviour of this system. The section on initial values plane

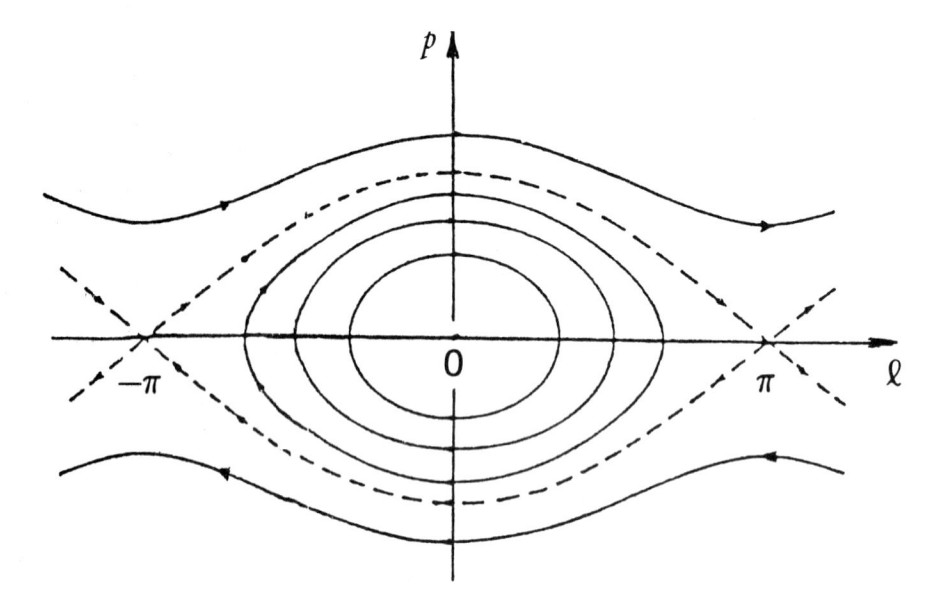

Figure 1. Trajectories of a mathematical pendulum; the broken line represents the separatrix.

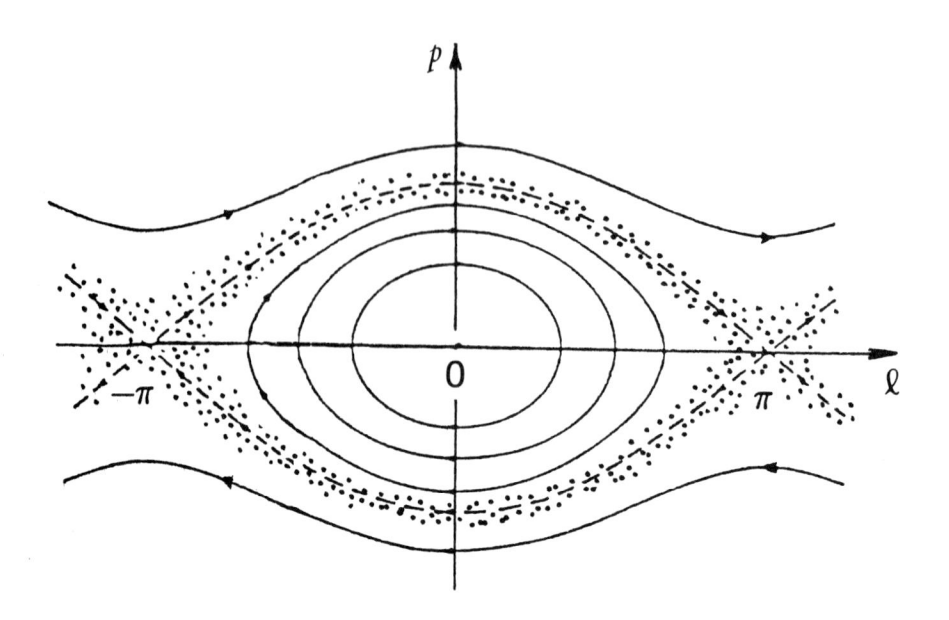

Figure 2. Destruction of the separatrix under the influence of a weak perturbation.

gives a selection of initial value maps which vividly illustrate this sensitivity. In the section on parameter space we present the results obtained by varying some parameters of the system. These parameter maps illustrate the variation of the total Hamiltonian with field strength. The final section summarises the properties of the described system and comments on the visual representation of the nonlinear mechanism that is responsible for the chaotic dynamics of the particle.

Analysis of the Motion

A general description of particle-field interaction illustrates one of the nondissipative processes that commonly features in plasma physics [Bird85]. The investigation of such motion is well handled within the framework of the Hamiltonian mechanics. The number of possible scenarios is large, and we restrict our attention here to particles entering the transverse magnetic field that is perturbed by a travelling plane wave, which is normal to this field.

Thus, when a charged particle moves in a constant transverse magnetic field its trajectory describes a circular orbit. The radius of the orbit is proportional to the initial velocity of the particle and its mass and is inversely proportional to its charge and the applied magnetic field. The plane of the orbit is perpendicular to the direction of the field and remains so at all times. If the particle's velocity is represented by a vector $\vec{v}$ and the magnetic field intensity is given by $\vec{B}$, and if the velocity of light is c, then the force $\vec{F}$ acting on a particle of charge e is

$$\vec{F} = e\frac{(\vec{v} \times \vec{B})}{c}$$

This is known as the magnetic field component of the Lorentz force and is perpendicular to both the velocity and field vectors. Its effect is continuous alteration of the velocity direction, thus introducing a rotation to the system.

On the other hand, in the presence of the linearly polarized electromagnetic wave, a charged particle oscillates. The polarization direction is identified with the electric field intensity vector $\vec{E}$, and the resultant force is given by

$$\vec{F} = e\vec{E}$$

The acceleration of the particle is in the direction of the electric field vector as the particle moves with the wave.

Consider now a charged particle moving in a constant magnetic field, in the presence of a linearly polarized wave propagating in the x direction. Assume that the constant magnetic field $\vec{B}$ is parallel to the z-axis, and the electric field $\vec{E}$ is parallel to the y-axis. The Lorentz force $\vec{F}$ provides a way to find the description of motion of such a particle in the field and can be expressed as

$$\vec{F} = e\left(\vec{E} + \frac{(\vec{v} \times \vec{B})}{c}\right) \tag{5}$$

This equation is quite general and describes the relativistic motion of a charged particle in a given external field. To avoid complexity in expressions and thus elucidate the meaning of the ongoing processes, the velocity of light c and the mass of the charged particle are set to unity.

The relativistic Hamiltonian describing the interaction of a charged particle with an external field can be written as [Heit54]

$$H = \sqrt{1 + (\vec{p} - e\vec{A})^2} \qquad (6)$$

where $\vec{p}$ is the generalized momentum and $\vec{A}$ is the vector potential of the field. Bearing in mind the assumed direction of our fields yields the expression for the vector potential in the form

$$A = Bx + \alpha E \sin \beta \qquad (7)$$

where α is the perturbation parameter, $\beta = kx - \Omega t$, with $\vec{k}$ denoting the wave vector and Ω representing the frequency of perturbation. Substitution of the Hamiltonian Eq. (6) together with Eq. (7) into the canonical Eqs. (1) leads to the equations of motion of a charged particle in a field [Zasl87]

$$\frac{dx}{dt} = \frac{p}{H}$$

$$\frac{dp}{dt} = -\left(\frac{e^2}{H}\right)(B^2 x + \alpha EB(kx \cos \beta + \sin \beta)$$

$$+ k\alpha^2 E^2 \sin \beta \cos \beta) \qquad (8)$$

An inspection of this equation shows that it contains two mutually exclusive types of motion: rotation (proportional to B) and translation. Consequently, we are dealing with the competition between two different symmetries. Depending on the value of the parameters, either one of these predominates or the system executes a chaotic motion, unable to decide which symmetry to follow. The behaviour is difficult to describe; it is our aim to visualize the above equation and convey some of the complexities that a particle in the field may experience.

Determination of the Trajectory

Numerical solution of Eqs. (8) is quite straightforward, and there are a number of various routines that are available. Reasonably good results can be obtained using numerous derivatives of the Euler method. However, in view of the strong nonlinearities involved in particle-wave interaction, the more accurate although considerably slower Runge-Kutta method was selected here. Due to a generally faster execution, several multistep methods were considered, but the need for accurate starting values, existence of a large local discretization error and inherent numerical instability meant that eventually they were abandoned. Direct coding of Eqs. (8) adds large computational overheads, due to the frequent calls to evaluate the trigonometric functions. A considerable reduction in computation has been achieved by transforming this system into an equivalent one, which involved calculation of these functions using the difference method. The resultant system then took the form

$$\frac{dx}{dt} = \frac{p}{H}$$
$$\frac{dp}{dt} = \frac{(Bx + \alpha ES)(B + \alpha k EC)}{H}$$
$$\frac{dS}{dt} = \left(k\frac{dx}{dt} - \Omega\right)C$$
$$\frac{dC}{dt} = -\left(k\frac{dx}{dt} - \Omega\right)S \tag{9}$$

with suitable initial conditions.

The nature of the problem required careful selection of the integration step size. Too large a step introduced numerical instabilities, completely dominating the behaviour of the physical system; too small a step contributed to a buildup of roundoff errors and also involved unproductive use of resources. Numerous trials led to an optimal choice, which depended on the parameters involved.

A frequent feature of nonlinear phenomena is the occurrence of transient chaotic behaviour. This arises for some values of parameters but eventually settles into a (quasi) periodic motion. To eliminate the possibility of transient chaos appearing, the first 300 periods were suppressed in most of the numerical simulations. Examples of phase space in this section contain trajectories pertaining to several different parameters. These multiple trajectories provide a better insight by putting the effect in perspective. The Poincaré maps were sampled at the frequency of the perturbation and contain 3000 points. The phase velocity of the perturbation in all instances was equal to the speed of light. The closed curves in phase space represent cross sections of a torus whose time axis is perpendicular to the plane of the display.

Several complementary issues are addressed in this paper. There are a number of free parameters pertaining to the description of the system. Hence we initially set them to some arbitrary values and study the effect of varying one of them on the trajectory of a particle. This has been done to obtain Plate 26, where different trajectories correspond to different starting values for the momentum. It should be remarked at this point that the effect of the initial phase value has a dramatic influence on the final result. The ratio of the cyclotron frequency Ω_c to the frequency of the perturbation Ω is $1/3$ and establishes the overall triangular-like shape of this phase plot. The three smallest orbits reflect the regular period-3 subharmonic motion. Further increase in the initial value of the momentum changes the trajectory into a single continuous curve that approaches the central point from three positions. The point of nearest approach recedes as the starting momentum rises, until a value is reached where the trajectory is no longer smooth and displays an irregular trace, typifying a chaotic tendency. A slightly slower particle leads to the creation of numerous islands stretching along the disrupted trajectory.

In the preceding description the value of the perturbation parameter α was 0.1. A reduction of this parameter removes the chaotic orbit in Plate 26 and restores the trajectories to their more circular form. This is to be expected, as in the absence of the perturbation the orbits are circular. The final image represents a set of concentric circles, which show a degree of distortion as the initial value of the momentum decreases. Chaotic trajectories are not avoided at this low value of α (0.01) and indeed reappear as soon as the initial momentum exceeds a certain higher level.

Changes in the magnetic field $\vec{B}$ directly influence the cyclotron frequency Ω_c, hence the resultant symmetry alters. Pictorial representation is given in Plate 27. At a certain value of the magnetic field the shape of the trajectory resembles a somewhat distorted circle. As the field strength increases so does the distortion, until at some critical value the originally single curve disintegrates into three separate curves. Still further increase reduces the size of individual orbits, which eventually disintegrate into several smaller units. As shown in Plate 27, the initial range of phases available to a particle is now restricted to three different subranges. When the value of the field reaches a level at which the transition takes place, the boundary between the two motions appears fractal.

Initial Values Plane

In the last section we looked at phase space, plotting momentum variable p against phase x. In response to changes in a single value of the parameter, different trajectories were obtained. Under certain conditions, several of these could be displayed simultaneously without cluttering the final image. However, the number of different conditions was limited, and other means of seeing a larger range in the variation of the starting conditions are needed. This brings us to the concept of the initial values plane in phase space.

It is well known that chaotic systems have a property of very sensitive dependence on the initial conditions. The ability to display this dependence could provide us with further insight into the nature of associated nonlinearities. The initial conditions plane contains a range of values which are used to start the integration. After propagating the solutions for a certain (fixed) length of time, the value of the associated Hamiltonian is colour coded and placed at the point occupying the cell in the initial conditions plane. The resolution of this plane was set to 640 × 350 cells. The process is very computationally intensive, as it consists of 224,000 separate integrations of Eqs. (9). In the absence of the magnetic field $\vec{B}$ the corresponding pictorial description is limited to a number of horizontal bands, showing the relative independence of the Hamiltonian on phase while clearly displaying the prominent momentum dependence. Plate 28 shows the variation in the Hamiltonian function as the initial phase (horizontal axis) and the momentum (vertical axis) of charged particles vary. The perturbation parameter $\alpha = 0.05$ and the ratio of the cyclotron to travelling wave frequencies is $1/3$. In the absence of perturbation, the colour bands are circular, of uniform thickness, contain no further structure, and clearly embody the underlying rotational symmetry. A particle in the relativistic domain shows some distortion in the circular bands; these are more pronounced in the left part of the figure, when the particle enters the field with the smallest phase value. The Hamiltonian function for slow-moving particles seems to be influenced most and begins to show a marked divergence between different trajectories, in response to the variation of the initial conditions. Increasing the value of the perturbation parameter to 0.1 strongly alters the appearance of phase space. The bands, denoting the constant levels of the Hamiltonian function, acquire tentacle-like forms at the outer parts of the figure that correspond to the relativistic particles. The region of slower particles shows minimal structure; there is a degree of randomness in the levels of the Hamiltonian. Still further increase in $\alpha(0.2)$ shows that for the initial conditions found in most of Plate 29 there is little correlation between the adjacent neighbouring points and the value of the final Hamiltonian. The tentacles become more extended, thinner, showing a tendency to fill large parts of phase space. As the chaotic domain approaches, these boundaries present fine, filigree-like whiskers with irregular, fractal boundaries. The existence of such boundaries is readily confirmed by successive expansion of the relevant areas of phase space. The presence of self-similar structure in these large magnifications provides strong evidence for the chaotic behaviour in the system. These experiments further demonstrate that the onset of chaos depends on the initial conditions of the charged particles as well as on the forcing amplitude. There is no structure in the very central part of Plates 28 and 29, but this absence is only apparent and arises from the limited colour resolution. Enlargement of that section is seen in Plate 30, which clearly emphasizes the period-3 oscillations.

The effect of the perturbation frequency has a very dramatic influence on the overall behaviour of the particle. In Plate 31 we see the result of varying the initial conditions, while keeping all the parameters underlying Plate 30 and using a higher frequency. There is a clear fourth-order symmetry, with a rich structure in the intermediate regions. Transitions between regular patterns are accompanied by whisker-like regions with high sensitivity to initial conditions. Increasing the perturbation amplitude destroys most of the regular features, and the transition regions become wider.

Parameter Space

The last section illustrated the effect of the change in the initial conditions on the Hamiltonian function. Now we examine a complementary effect, namely the influence of the perturbation strength on the final result. In the following plates the perturbation parameter α is increasing in the direction of the horizontal axis, and the momentum changes along the vertical axis. During the sweep of the range of the perturbation parameter, each pixel in the final image corresponds to a different set of starting values. In Plate 32 the parameter α varies in the range (-0.3, 0.3). Slow particles are at the bottom of that figure, and their velocity increases upwards. It is seen that a wide central strip is clear of any irregularities, and variation in the particle's momentum produces no appreciable results. This situation changes as the momentum increases, as can be seen in Plate 33, where only the positive part of α has been displayed. Once again, a small area around the left of the figure is devoid of any structure, but this distance decreases for faster particles. An expansion of the relativistic region, and inclusion of negative electric field amplitudes as well, is shown in Plate 34, where the splitting of a colour band into a number of fine filaments is clearly visible. Further increase in the strength of the perturbation completely destroys these, and the loss of order predominates. It is often found that enlarging the chaotic region brings up islands of regular motion surrounded by regions of disorder. A small portion of Plate 32 has been enlarged; this property is clearly seen in Plate 35.

Conclusion

The main aim of this paper was to visualize the chaotic dynamics experienced by a charged particle in a magnetic field, subject to a disturbance caused by a travelling plane wave. Such a system belongs to the group of Hamiltonian systems, which represent nondissipative processes. As there are no losses, it makes no sense to talk about attractors. This is fully supported by presenting a Poincaré map, which tends to fill up large regions of phase space with uniform probability, in contrast to the clustering of points in fractal shapes found when dissipation is present.

From the pictorial presentation it can be affirmed that in the relativistic domain the dynamics of the particles becomes chaotic for very small values of the perturbation parameter. At the other extreme, slow particles exhibit a regular motion confined to islands that, for stronger perturbation, are separated by stochastic regions.

When slow electromagnetic waves (phase velocity is less than the speed of light) are present, the charged particle can become trapped by the wave and, further, experience regular acceleration. Due to the constraints imposed by the nature of the problem, this does not occur in the case under consideration here when the phase velocity is equal to the speed of light. However, it is found that for any value of the perturbation parameter there exists a value of the initial momentum of the particle, above which the particle's dynamics becomes chaotic. This is represented by a mixing taking place in phase space. The condition that determines the onset of this chaotic behaviour is known [Zasl87].

A small perturbation parameter gives rise to stochastic bands which are separated by invariant tori. Increasing the momentum of the particle destroys these invariant curves, and a random-like motion takes place where stochastic acceleration of particles occurs. Due to the interaction of particles with the field, the nonlinear processes in this regime provide a mechanism for damping of the waves.

The total Hamiltonian in Eq. (2) consists of two parts: H_0 and V. The former gives rise to the invariant tori throughout phase space; the latter destroys these curves and replaces them with thin channels, where the dynamics of particles is chaotic. These channels provide the means for a particle to move without restriction around phase space, thus changing, without a bound, the associated invariant quantities. The particles inside these channels can diffuse; others are bound to execute regular oscillations inside the stable islands. As the energy of the particle increases the size of the channels diminishes, that is, the value of the distribution function decreases for fast particles.

Initial conditions maps provide a global view of the system by displaying its behaviour for a range of starting conditions. They provide a measure of the dynamical distance of the system from the steady state, in the absence of perturbation. In the vicinity of the chaotic regions the highly mixed whiskers indicate the presence of fractal boundaries. The originally smooth boundaries now become diffuse, encompassing most of the available space. In this regime it is not possible to predict the final state of the system if there are any inaccuracies in the initial conditions.

The computation connected with the above experimentation demanded a repetitive evaluation of a large number of identical processes with differing sets of data. Consequently, in view of its nature such calculations are ideal for parallel processing. Indeed, some of the results were thus obtained, resulting in considerable time gains.

Acknowledgements. I would like to thank D. Trenaman for an excellent preparation of colour slides.

REFERENCES

[Moon87]
   Moon, F.C., *Chaotic Vibrations*, New York: John Wiley and Sons, 1987.
[Arno63]
   Arnol'd, V.I., Small denominators and problems of stability of motion in classical and celestial mechanics, *Russian Math. Surveys*, Vol. 18, p. 85, 1963.
[Bird85]
   Birdsall, C.K., and Langdon, A.B., *Plasma Physics Via Computer Simulation*, New York: McGraw-Hill, 1985.
[Heit54]
   Heitler, W., *The Quantum Theory of Radiation*, Oxford, UK: Oxford University Press, 1954.
[Walk69]
   Walker, G.H., and Ford, J., Amplitude instability and ergodic behaviour for conservative nonlinear oscillator systems, *Phys. Rev.*, Vol. 188, p. 416, 1969.

[Zasl87]
Zaslavskii, G.M., Natenzon, M.Y., Petrovichev, B.A., Sagdeev, R.Z., and Chernikov, A.A., Stochastic acceleration of relativistic particles in a magnetic field, *Sov. Phys. JETP*, Vol. 66, p. 496, 1987.

Chaos in Physical Systems

Tom Mullin

Abstract

It is now an established fact that simple dynamical systems governed by deterministic equations can exhibit irregular or chaotic behaviour with well prescribed input conditions. We illustrate the basic principles involved in modern thinking on chaos in dynamical phenomena with two particular examples, the simple pendulum and a nonlinear electronic oscillator. The application of these ideas to the origins of fluid flow turbulence is then discussed. We show that, in certain situations, ideas from finite dimensional dynamical systems do aid the understanding of the transition to turbulence.

Introduction

The field of classical dynamics usually involves the study of a system governed by some form of Newton's equations of motion. The general consensus of opinion is that even very complicated dynamical problems will be completely solvable by the use of large computers at some finite time in the future, e.g., in aerodynamics we are often assured that the computer will 'soon' replace the wind tunnel. However, it is now realised that simple finite dimensional systems governed by deterministic equations (i.e., future events are determined by present conditions and a governing set of 'rules') can exhibit chaotic or irregular behaviour with well defined input conditions. The chaos does not arise because of imprecision or noise at the input but is a feature of the system itself. It may be thought, therefore, that a complete description of the behaviour of a nonlinear system will never be possible. However, the appearance of chaos in many systems follows distinctive paths which are rich in structure. We will illustrate this point by use of three physical examples which all follow the same route to chaos.

One Route to Chaos: Period Doubling

The transition to chaotic motion is often found to follow well defined paths which can sometimes be displayed by extremely simple equations. An example of one of these equations is the so-called quadratic map, where an iterative relationship has a quadratic form. Specifically

$$x_{n+1} = 4Lx_n(1 - x_n) \tag{1}$$

where x_n is the nth iterate of the relationship, and L is some parameter in the range between 0 and 1. Clearly, 0 satisfies Eq. (1), and 1 also gives 0 as the final outcome. However, any other starting value between 0 and 1 gives a nonzero answer, provided $L > 0.25$. If L is in the range $0.25 < L < 0.75$ then we obtain a unique outcome for any starting value other than 0 or 1. This we call a fixed point of the map.

If L is now increased to 0.785, the iteration scheme no longer converges to a fixed point but instead oscillates between two values. We call this a 2-cycle. Further increase in L leads to a breakdown of this 2-cycle, to be replaced by one of period 4, and then period 8, 16, etc., with yet further increase in L. Each new cycle appears at smaller and smaller increments of the parameter L. Feigenbaum [Feig80] showed that the ratio of these critical values for the appearance of new cycles has a geometrical relationship. This means that an accumulation point is rapidly reached where essentially chaotic iteration breaks out. A very important feature of Feigenbaum's discovery is that this relationship is universal, being independent of the details of the mapping. Thus, if a period doubling sequence is found in an equation or a physical experiment then it should by expected to follow this rule. We show a plot of the final 100 iterates of such a mapping as a function of L in Figure 1.

A Mechanical System

An example of chaos in a simple system is given by the parametrically excited pendulum. In this arrangement, the pivot point of a simple rigid pendulum is moved vertically up and down in a sinusoidal manner. The equation of motion is given by

$$\frac{d^2 A}{dt^2} + D\frac{dA}{dt} + (1 + B\omega^2 \cos \omega t) \sin A = 0 \tag{2}$$

where A is the angle of swing of the pendulum, D is the damping coefficient, B is the nondimensional amplitude of the applied oscillation, and ω is the ratio of the applied frequency to the natural frequency. Equation (2) is satisfied for all values of B and ω by $A = 0$, but it becomes unstable at certain values of these parameters. Physically, this corresponds to simple up and down motion of the pendulum being replaced by swinging from side to side and is known as parametric excitation of the pendulum. This phenomenon is well known and was first discussed by Lord Rayleigh [Rayl02].

The principal region of instability contains motion of the pendulum at half the driving frequency. However, there are other ranges of B and ω within the principal band where the simple periodic motion gives way to chaotic oscillations; thus the long term motion is no longer predictable. This is particularly

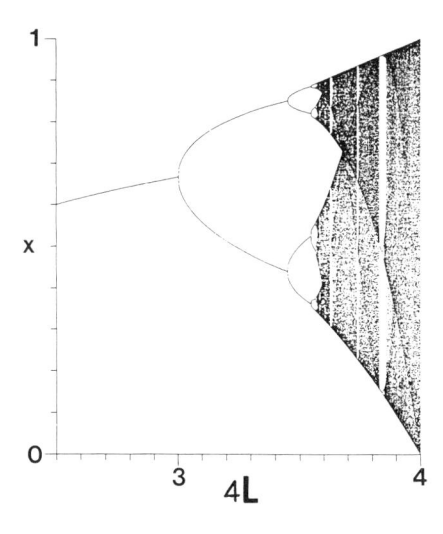

Figure 1. Plot of the final 100 iterates of the quadratic map as a function of $4L$.

evident in the physical system if the pendulum is allowed to rotate through 360°. Here the direction of rotation, as well as the amplitude, changes in an apparently random fashion. Chaos arises through a period doubling sequence and was first discovered in this problem by Leven et al. [Leve85]. The unpredictable motion occurs in both the numerical integration of Eq. (2), which can be performed on a microcomputer, and in the physical system. Thus, we are forced to the conclusion that it is a fundamental property of the system and is not due to imprecision or randomness in either realisation.

An Electronic System

A convenient demonstration of the period doubling route to chaos is provided by the nonlinear electronic oscillator shown schematically in Figure 2. The additional elements a and b added to the basic oscillator circuit are admittances. They allow us to explore a wide range of parameter space. For the present discussion we consider b to be fixed; we vary a to explore the period doubling cascade.

The output from the oscillator is displayed in the form of phase portrait diagrams in Figure 3. These show the temporal structure of the output at various set values of a. The simplest behaviour is shown in Figure 3a and corresponds to a sinusoidal oscillation in the circuit. In this representation one orbit around the loop corresponds to the period of the oscillation.

Figure 3b shows that above a critical value of a the cycle repeats every two orbits, i.e., the basic period doubles. It should be noted that the phase portrait is no longer planar but is three-dimensional, otherwise the orbits would self intersect. Further change in a leads to the period 4 orbit of Figure 3c, and yet further increase in a gives rise to the chaotic trajectory shown in Figure 3d. With this present circuit it is only possible to observe the first few period doublings

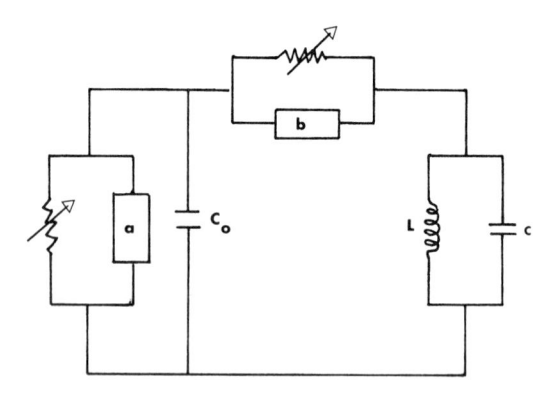

Figure 2. Schematic diagram of the nonlinear oscillator circuit.

(up to period 8), because the critical points approach each other geometrically (as discovered by Feigenbaum [Feig80]). Thus, any small amount of noise in the circuitry overcomes the delicate features near the end of the cascade and prevents observation of larger period orbits.

The circuit thus provides a good example of a nonlinear feedback network, where chaos arises by a mechanism which can be understood from very simple sets of equations. This general principle could apply to any nonlinear control system where, in general, one should expect to encounter chaos.

Hydrodynamics

The main thrust of our research effort in the Nonlinear Systems Group in Oxford is directed towards the understanding of the motion of fluids, and in particular the appearance of turbulence. Perhaps the simplest example which encapsulates the problem is given by the flow from a kitchen tap. If the tap is opened very slightly, then the water falls in a smooth jet; we call this laminar flow. It is not a very common flow state in nature, yet it still tests the limits of the largest computers when calculating them from the established equations of motion, the Navier Stokes equations.

If the tap is now opened further, then the jet becomes rough; we call this turbulent flow. This is the most common type of flow, but it is not possible to calculate such flows starting from the basic equations of motion. It is, of course, possible to make progress in many practical situations by building *ad hoc* models which capture some of the flow features. However, if we wish to progress beyond this stage, then one way to proceed would be to increase our knowledge of the fundamental processes involved in the onset of turbulence in a fluid.

A classical problem which displays the onset of irregular motion through a sequence of critical events is afforded by Taylor-Couette flow, first studied by G.I. Taylor [Tayl23]. In this problem we consider the motion in the fluid between

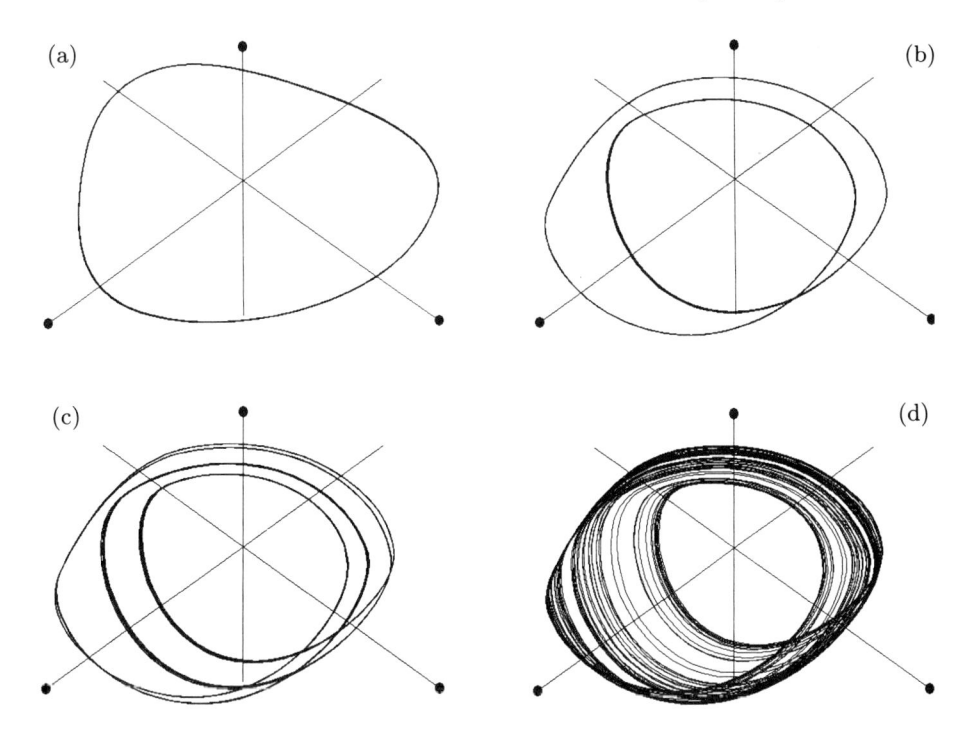

Figure 3. Phase portraits of the output from the electronic oscillator. (a) Period 1; (b) period 2; (c) period 4; (d) chaos.

two concentric rotating cylinders, where the inner rotates and the outer is held stationary. At small values of the nondimensionalised speed, or Reynolds number Re, the motion is mainly in circles around the inner cylinder except for three-dimensional motion near the ends. When Re is increased beyond a critical range secondary motion sets in, which takes the form of a set of cells stacked upon each other, as shown in Figure 4. The main azimuthal motion now has a secondary circular motion superimposed in an orthogonal plane, so that particles in the flow follow spiral paths.

As Re increases, further waves appear on the interfaces of the cells, as shown in Figure 5. If we were to take measurements at a single point in the flow field in this state, using a laser probe for example, then the signal obtained would take the form of a simple oscillation.

Further increase in Re leads to a more complicated motion, until eventually the flow within each cell becomes chaotic in the temporal sense; but the overall spatial cellular structure is maintained at this stage. Therefore, it is not clear whether this can be described as turbulent motion where, generally, both temporal and spatial disorder are found.

The theoretical model which is often used to describe the events in this so-called Taylor-Couette flow considers the cylinders to be infinitely long. This

Figure 4. Front view of Taylor Cells. Small light reflecting plates have been added to the flow for visualization.

allows considerable simplification of the equations of motion, for it allows the basic state to consist of purely azimuthal motion and thus ignores the inevitable three-dimensional effects at the ends in the physical system. Further, the cellular motion can be represented by trigonometric functions. This mathematical abstraction has guided much of the past experimental work; consequently, most of the systems are very long in an attempt to 'minimise' the finite length effects.

Important work by Benjamin [Benj78] showed, however, that the abstract situation can never be approached in practice, and that there are important qualitative differences between the theoretical model and the experiment. One important practical consequence of using a large scale experimental arrangement is that many different solutions to the problem can coexist at the same values of the control parameters. This multiplicity of states is a common feature of nonlinear systems and may be thought of in the following way. Suppose there is a landscape with many dips in the surface and that a ball is cast onto the surface. The ball may land in any of the dips and remain there unless it is perturbed. Each dip is analogous to a different solution to our problem. If the system is large and hence the solution set is also very large, external fluctuations may have an overriding effect on the dynamics so that the system will never attain equilibrium. Thus, uncontrollable external fluctuations may mask the fundamental processes involved in the production of chaos.

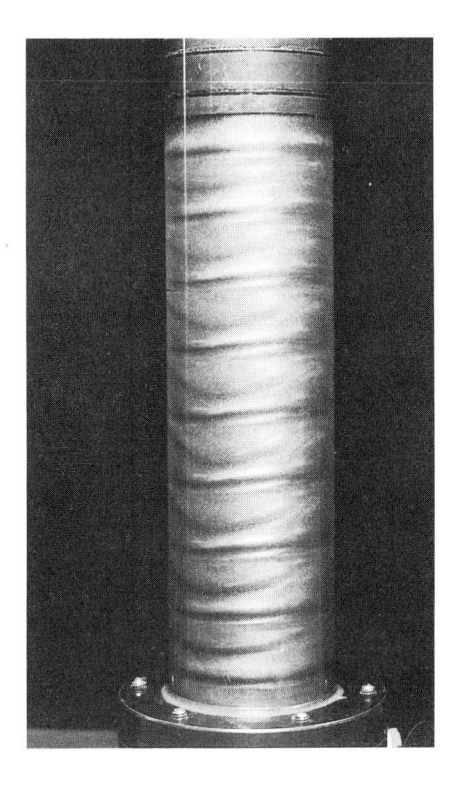

Figure 5. Wavy Taylor vortices.

In view of these difficulties, we decided to investigate small scale systems where the number of solutions is limited and fundamental interactions are more readily studied. The experimental work still requires stringent control on all the parameters, since many of the observed events are delicate and precision is required for repeatable results.

We present in Figure 6 the phase portraits for a period doubling cascade obtained in Taylor-Couette flow with four cells. These phase portraits, which show period 1, 2, 4 and chaos, have been reconstructed from the time series obtained from a laser Doppler velocimeter [Darb91] which measures the radial velocity component at a point in the flow. The period doubling sequence rides on top of a wavy flow such as the one shown in Figure 5, but we have filtered out the higher frequency ($\approx 8\times$ period 1) associated with this wave for ease of presentation. Thus the basic motion is not a simple limit cycle, as in the case of the electronic oscillator, but is really a torus in phase space. We would therefore require a minimum of four dimensions to display the remainder of the sequence.

Many other forms of dynamical phenomena which have their roots in finite dimensional dynamical systems have now been discovered in this flow [Mull89a; Mull89b]. These results give hope that certain features of fluid flows may be describable by simple sets of equations which can be solved on microcomputers. However, it must be remembered that we are only attempting to describe

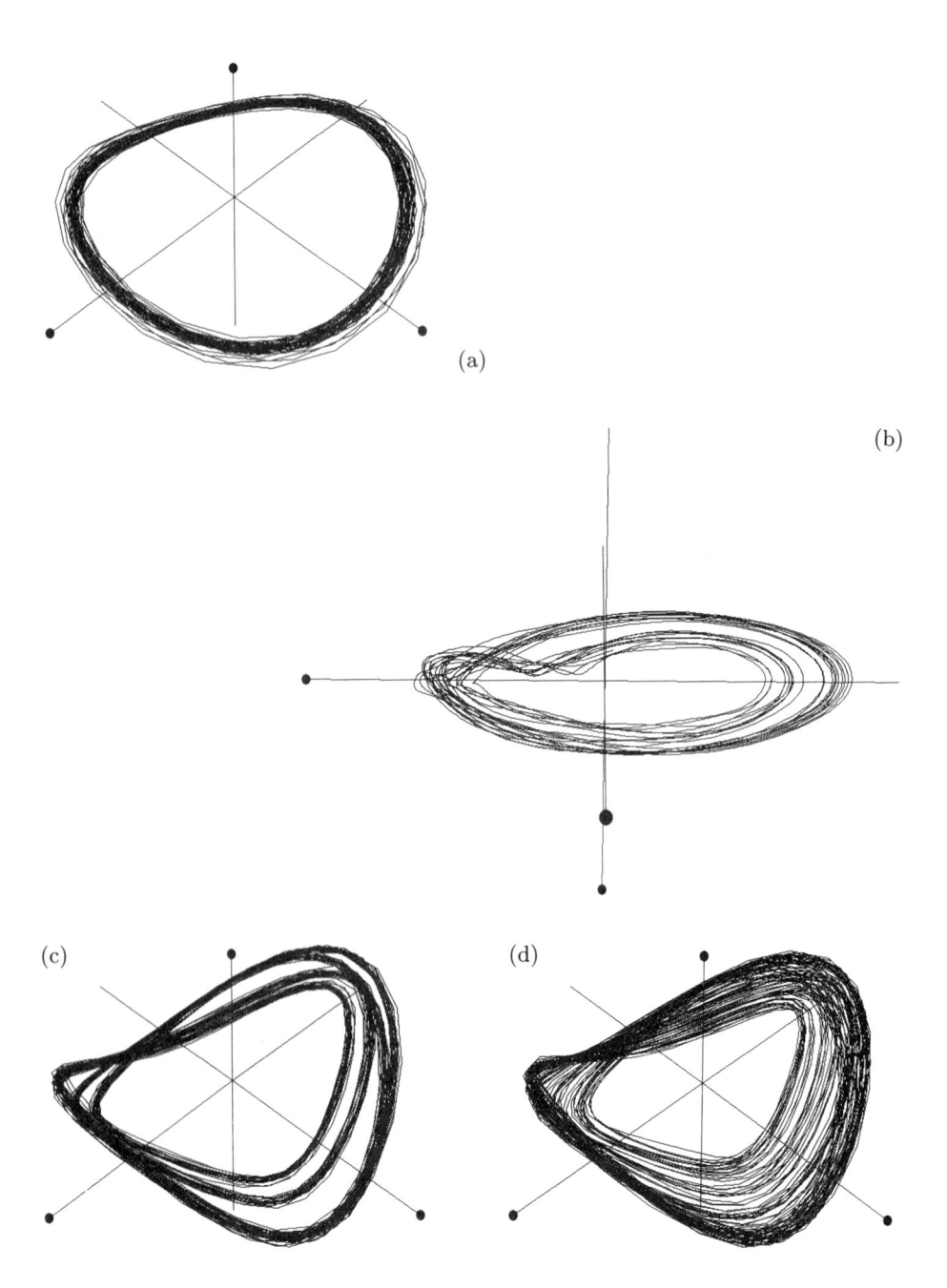

Figure 6. Phase portraits obtained from Taylor vortex flow. (a) Period 1; (b) period 2; (c) period 4; (d) chaos.

the temporal evolution of a particular and somewhat special flow and that the problem of turbulence is still far beyond our grasp.

Acknowledgements. I am grateful to Anne Skeldon and Jonathan Healey for allowing me to quote from their work on the pendulum and electronic oscillator, respectively. Drs. David Broomhead and Robin Jones of RSRE Malvern gave much appreciated advice on signal processing and provided the electronic oscillator. Finally, I would like to thank the SERC which supports our work through the Nonlinear Initiative.

REFERENCES

[Benj78]
Benjamin, T.B., Bifurcation phenomena in steady flows of a viscous liquid. Parts I (Theory) and II (Experiments), *Proc. Roy. Soc. London Ser. A*, Vol. 359, pp. 1, 27, 1978.

[Darb91]
Darbyshire, A.G., and Price, T.J., Phase portraits from chaotic time series, *this volume*, pp. 247—257, 1991.

[Feig80]
Feigenbaum, M.J., Universal behaviour in nonlinear systems, *Los Alamos Science*, Vol. 1, p. 4, 1980.

[Leve85]
Leven, R.W., Pompe, B., Wilke, C., and Koch, B.P., Experiments on periodic and chaotic motions of a parametrically forced pendulum, *Physica*, Vol. 16D, p. 371, 1985.

[Mull89a]
Mullin, T., and Darbyshire, A.G., Intermittency in a rotating annular flow, *Europhys. Lett.*, Vol. 9, p. 669, 1989.

[Mull89b]
Mullin, T., and Price, T.J., An experimental observation of chaos arising from the interaction of steady and time dependent flows, *Nature*, Vol. 340, p. 294, 1989.

[Rayl02]
Rayleigh, Lord, *Scientific Papers*, Vol. III, pp. 1–14, Cambridge, UK: Cambridge Univ. Press, 1902.

[Tayl23]
Taylor, G.I., Stability of a viscous liquid contained between two rotating cylinders, *Phil. Trans. Roy. Soc. A*, Vol. 223, p. 289, 1923.

Phase Portraits from Chaotic Time Series

A.G. Darbyshire and T.J. Price

Abstract

The reconstruction of a phase space using the method of delays and singular value decomposition is discussed. These procedures are applied to data from both experiments and numerical simulations. It is then shown how computer graphics techniques may be employed in the representation of phase portraits to clarify their topological structure.

Introduction

It is now well established that simple low-dimensional dynamical systems can display chaotic and other complicated behaviour [Lore63]. Also, physical systems can, in certain circumstances, appear to show similar kinds of phenomena [Schu84]. In order that comparison can be made between theory and experiment, it is necessary to describe them both within a common framework. In this paper we firstly introduce terminology and some of the ideas of phase or state space analysis in the context of dynamical systems. We then show how, in general, a phase portrait can be constructed from an experimental time series. In the third section we show practical examples of the reconstruction method, including the effect of noise. The final part of the paper is concerned with the presentation of the resulting phase portraits, using techniques which give an understandable three-dimensional representation on a two-dimensional surface, such as a computer graphics screen.

Phase Spaces

Formally, an n component system which evolves as a function of time can be described at any time t by a set of values, $\mathbf{P} = (p_1, p_2, \ldots, p_n)$, and their time derivatives $\mathbf{P}' = (p_1', p_2', \ldots, p_n')$. Thus the state of the system is given by $(\mathbf{P}(t), \mathbf{P}'(t))$, which is a point in a space of $2n$ dimensions. This space is called phase space. As time evolves, $(\mathbf{P}(t), \mathbf{P}'(t))$ traces out a path in this space called the trajectory. The phase portrait is just this trajectory allowed to build up over a considerable time. The distribution of the points in phase space gives statistical information, and the flow between points contains dynamical information. Sometimes, however, it is easier to think of phase space as containing all the degrees of freedom of a particular system.

Consider the example of a simple pendulum performing small amplitude oscillations. The equation of motion is a second-order differential equation and can therefore be described in two dimensions, corresponding to the position and velocity of the pendulum. The phase portraits are shown in Figure 1 for two cases:

undamped motion;
damped motion.

In the undamped case the trajectory is just a circle, where each revolution of the circle corresponds to one complete cycle of the pendulum. In the damped case the phase portrait is a spiral, reflecting the amplitude of the oscillations as they decay towards a fixed point at the origin, where the pendulum will come to rest. Such a system is not chaotic.

Let us look at a simple dynamical system which exhibits chaos. The Rossler equations are [Ross76]

$$\frac{dx}{dt} = -1(z+y)$$

$$\frac{dy}{dt} = x + ay$$

$$\frac{dz}{dt} = b + z(x-c)$$

Figure 2 shows the trajectory in (x, y, z) space produced by integrating the equations with the parameter values $a = 0.2$, $b = 0.2$, $c = 4.6$. Packard et al. [Pack80]

(a)   (b)

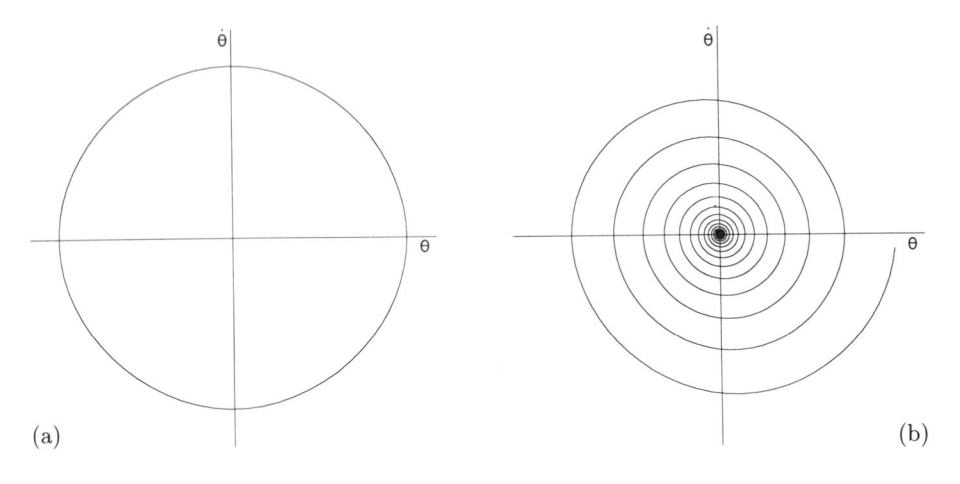

Figure 1. Phase portrait of simple pendulum. (a) Undamped; (b) damped.

showed that it was possible to produce a trajectory using, for instance, x and its first and second time derivatives, which appeared to reflect the dynamics of the original attractor. In the context of experimental work this is important, since it suggests that from a single time series it is possible to construct a multi-dimensional representation of the system, rather than having to use multiple time series as was implied in the production of the phase portraits in both Figures 1 and 2. Independently, Takens [Take81] developed a similar approach, known as the 'method of delays', and also placed such techniques within a mathematical context.

Time Series Analysis

A time series is a set of values sampled at a regular period; let this period be called the sample time t_S. The whole series is a vector $\mathbf{V} = (v_1, v_2, \ldots, v_N)$. The simplest implementation of the method of delays is to use n consecutive samples to create an n-dimensional embedding space. Therefore, if we set $n = 3$, a trajectory would be built up using

$$(v_i, v_{i+1}, v_{i+2}) \qquad 1 \leq i \leq N - 2$$

However, if the signal is sampled at a fast enough rate, such that the discrete nature of the trajectory is not readily apparent, the samples within the n-window become highly correlated, and the phase portrait lies close to the $(1, 1, 1 \ldots)$ diagonal in the embedding space. To overcome this difficulty it is usual to specify a lag time $t_L = j t_s$, where $j > 1$. Hence, in an $n = 3$ embedding space the trajectory would be built up using

$$(v_i, v_{i+j}, v_{i+2j}) \qquad 1 \leq i \leq N - 2j$$

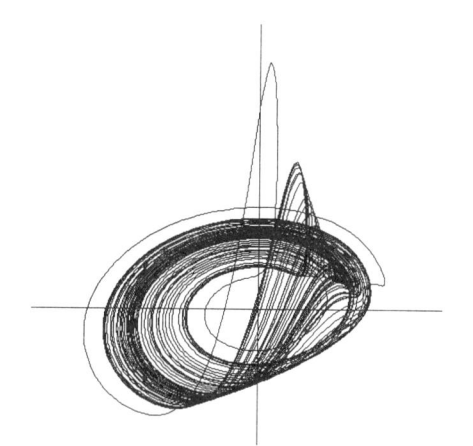

Figure 2. Phase portrait produced by integration of the Rossler equations.

As j is increased the components become more statistically independent. Figure 3 shows the Rossler attractor in an $n = 3$ embedding space, reconstructed from the time series of the x coordinate using the method of delays with a lag time of 10 samples. The problems with the method of delays are threefold. One is common to all reconstruction methods, *a priori*, how does one know the size of the embedding space in which to locate the phase portrait? The second difficulty lies in the optimum choice of t_L. The final problem concerns the effect of noise present in the signal.

Figure 4a is a time series from a well-controlled Taylor-Couette flow experiment [Mull90]. It shows a strong periodic signal and also a noise component with an amplitude approximately ten percent that of the main oscillation. The reconstructed phase portrait, using the method of delays, is shown in Figure 4b, where the noise component leads to a broadening of the trajectory. Comparison with the previous reconstruction of the 'noiseless' numerical data illustrates the effect of the noise. In an attempt to overcome these difficulties, Broomhead and King [Broo86] developed a statistical approach to the method of delays based on singular value decomposition (SVD), which is a well known technique in linear algebra [Golu83].

The singular systems approach involves building up a trajectory matrix $[X]$, which contains the n-dimensional vectors used for the method of delays with $j = 1$. Broomhead and King show that by computing the SVD of the covariance matrix $[X]^\mathrm{T}[X]$ it is possible to obtain a set of eigenvectors which are the orthogonal singular vectors, and also the singular values which are the square roots of the corresponding eigenvalues. The SVD of the covariance matrix is calculated by diagonalizing $[X]^\mathrm{T}[X]$.

Broomhead and King suggest that to help envisage these vectors one thinks of the trajectory exploring an n-dimensional ellipsoid; then the singular vectors

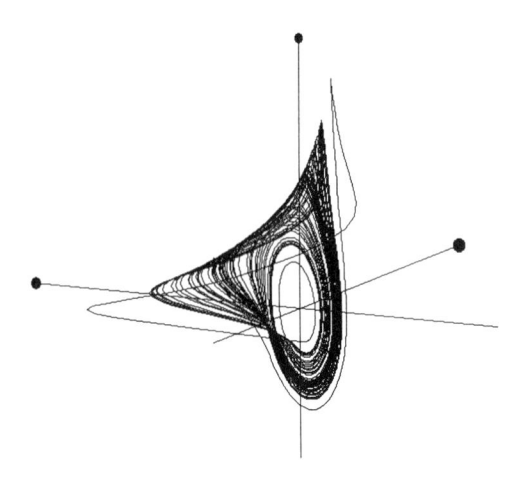

Figure 3. Reconstruction of the Rossler attractor using the method of delays.

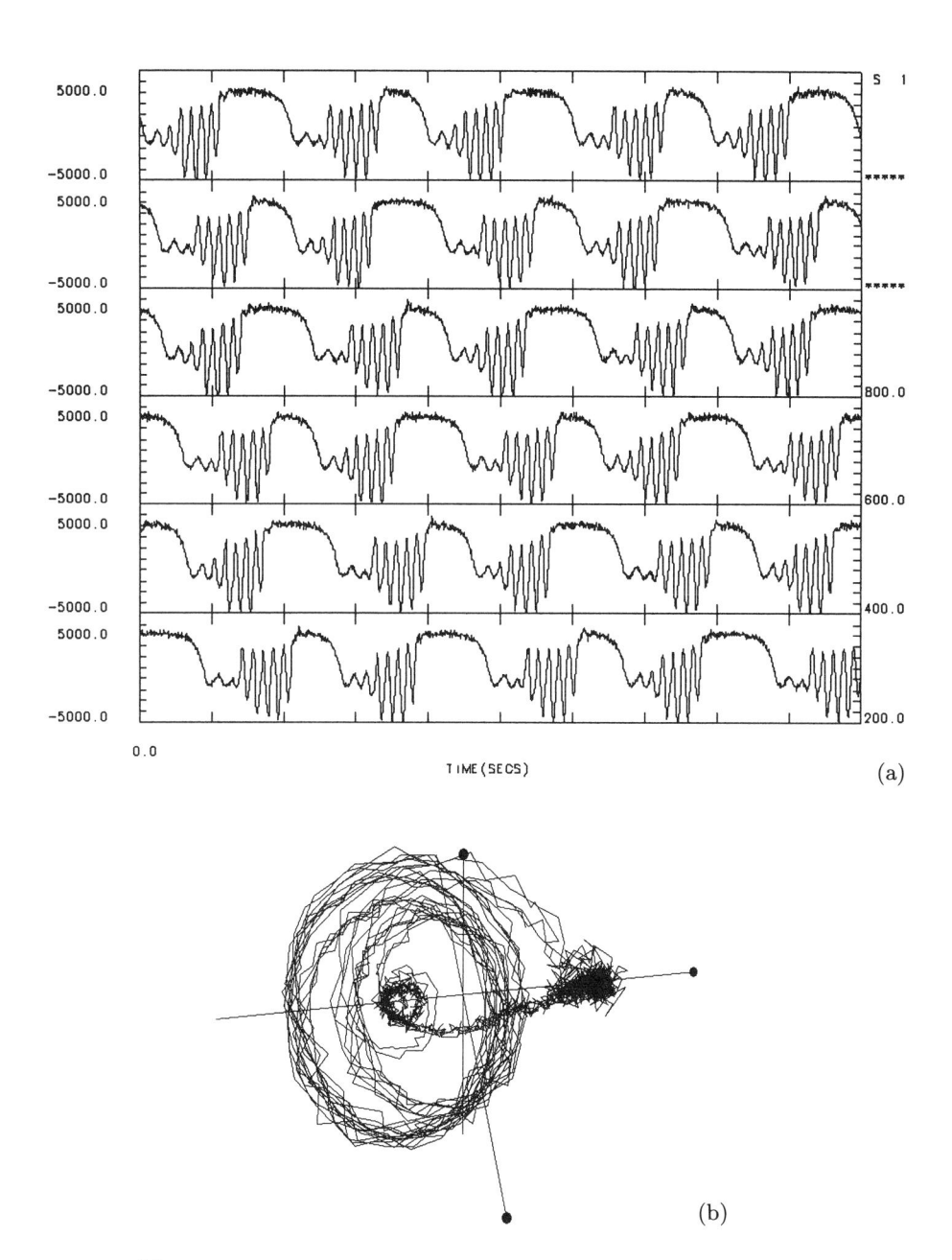

Figure 4. (a) Velocity time series from the Taylor-Couette experiment; (b) phase space reconstruction using the method of delays.

and values give the directions and lengths of the principal axes. The number of nonzero singular values is less than or equal to n. This gives an upper limit on the embedding dimension in this case. When noise is present, all the singular values are greater than zero; therefore, it might be suggested that nothing is gained in respect to the size of the embedding space needed to describe the signal. However, the isotropic noise component tends to dominate at small or vanishing singular values, since it adds equally to all the directions, and hence the singular value spectrum contains a noise floor. The singular values corresponding to the deterministic part of the signal will be prominent above the noise floor: thus, an upper limit can still be placed on the number of significant dimensions. Figure 5a is a plot of the normalized singular value spectrum for the time series used to produce Figure 4b plotted on a logarithmic scale. The noise floor and the singular values due to the periodic signal are clearly visible. The time series is then projected onto the significant singular vectors to produce the phase portrait shown in Figure 5b. When compared with the portrait produced by the method of delays, both show the same general structure. But the singular value approach is less susceptible to the influence of noise.

Graphical Presentation of Phase Portraits

The method of singular value decomposition takes a set of N scalar values sampled at regular intervals from a time series obtained from a physical system or a theoretical model. It generates an ordered array of N vectors, each of which specifies a point in phase space. This sequence of points, by its discrete nature, does not uniquely define a trajectory representing the continuous time evolution of the system. Nevertheless, if the data sampling rate is sufficiently great in comparison

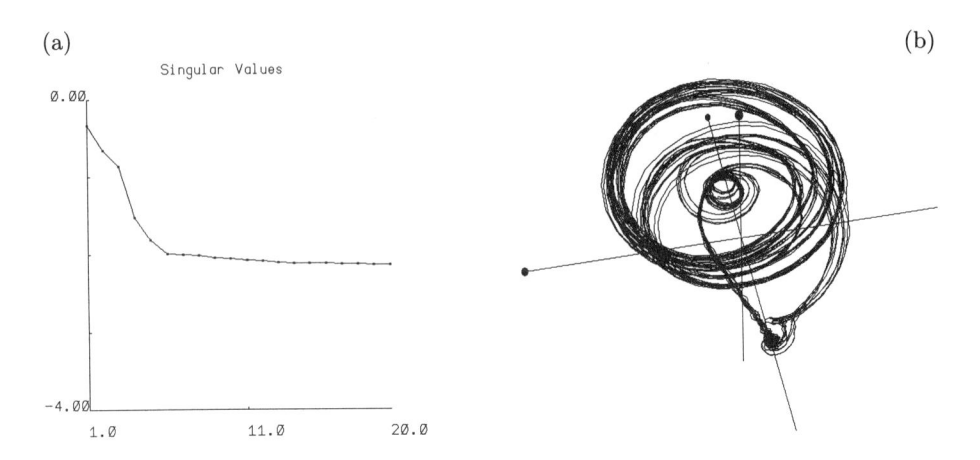

Figure 5. (a) Normalised singular value spectrum plotted on the logarithmic scale; (b) phase portrait reconstruction using singular value decomposition.

to the highest significant frequency present in the signal (an oversampling factor in excess of 50:1 is desirable) then, in the absence of noise, the error between the true trajectory and a piecewise linear interpolation between successive points is acceptably small. In practice, each point has an error footprint in phase space, resulting from quantisation, measurement and system noise.

The oversampling factor can be reduced to 10:1 if higher-order curve fitting is used. Casdagli [Casd89] compares the relative merits of radial basis functions, global polynomial fitting and local splines. However, in applications involving the manipulation and display of phase portraits, the reduction in the number of data points is outweighed by the increased complexity of interpolation. On the other hand, curve fitting is worthwhile if quantitative measurements of the trajectory's geometry are to be made, for example the calculation of Lyapunov exponents. Such numerical processing of trajectories is not discussed here. We concentrate instead on the problem of revealing the qualitative structure of phase portraits.

It is often necessary to discover the manner in which the topological form of a phase portrait evolves as some system control parameter is varied. For example, a fixed point may grow into a closed loop, which subsequently expands into a torus. In order to make such qualitative judgments it is necessary to have some means of representing the trajectory visually; surprisingly, little attention has hitherto been paid to this aspect of phase portrait interpretation.

Graphical display devices are grouped broadly into two categories, raster and vector plotters. The task of displaying a series of connected line segments is ideally suited to the nature of a vector device. The simplest procedure performs an axonometric orthographic projection of the trajectory onto a plane. This yields a set of (x, y) coordinates to be joined sequentially. Figure 6 shows a phase portrait obtained from a set of three ordinary differential equations (it therefore occupies a three-dimensional phase space) after a two-dimensional orthographic projection.

The main disadvantage of this method is that all information relating to dimensions orthogonal to the projection plane is lost. At the very least, an attempt

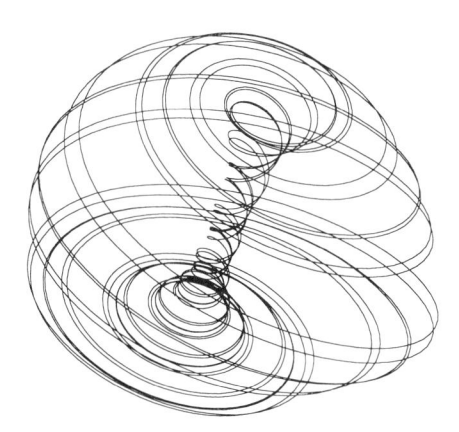

Figure 6. Axonometric orthographic projection of a three-dimensional phase portrait.

should be made to incorporate z-axis information into the image. One basic solution is depth cueing, whereby the intensity of the trajectory is modulated so that the nearest lines are plotted most brightly. This system can be implemented by sending three components of the trajectory position vectors to three D/A channels in order to control the X and Y plates and the Z modulation input of an oscilloscope.

If depth cueing is not available, the phase portrait can be rotated in real time to give the impression of a three-dimensional structure from a moving two-dimensional projection. The viewer can extract the z-axis information from the relative velocities of different parts of the image, but there remains an ambiguity (exemplified by the Necker cube) between back and front, so that the whole picture appears to flip inside out. Depth cueing is free from this shortcoming, but of course neither method can be used to produce black and white hard copies (e.g., for publication of results) that convey more than two-dimensional information.

It is more common for computers to be equipped with raster displays, but this by no means precludes the use of the above techniques. Furthermore, other options for three-dimensional portrayal are available which are not possible on a vector device; some of these are discussed below.

The task of interpreting the full structure of a three-dimensional phase portrait from its two-dimensional projections is analogous to ascertaining the shape of a tangle of string from the shadow it casts, where the shadow may be thrown in an arbitrary direction. Then, just as we would prefer to view the tangle of string directly, it would be advantageous if a computer could render a stored representation of a trajectory as if it were a physical object. Although the standard methods of hidden-surface removal and shading according to light sources are well known [Fole82], these techniques cannot be directly applied to the phase portrait because it is composed of line segments, not planar polygons. Hidden-surface removal relies on the existence of objects of nonzero projected area to block the view of other, more distant objects, and shading requires a surface in order to define a normal vector. A line meets neither of these criteria.

One solution (taking a cue from the string analogy) is to thicken the line so that it becomes circular in cross-section. However, the rendering of cylinders is computationally expensive, so we have chosen an alternative method that also provides extra visual information by which to judge the orientation of a trajectory in phase space.

To define unambiguously a surface normal at each point on a trajectory, it is possible to make use of the instantaneous centre of curvature at that point. The position of the centre of curvature relative to the trajectory yields a vector whose direction can be equated with the surface normal. Now the trajectory can be represented as a ribbon of constant nonzero width, which is amenable to efficient rendering since just one polygon need be plotted in place of each line segment, for which the cylinder representation requires at least twenty polygons. The ribbon's projected width—the ribbon appears widest when face-on and thinnest when edge-on to the viewer—supplies additional information about its orientation that could not be obtained from a similar view of a cylindrical trajectory. In Plate 36 the ribbon method has been applied to the data set of Figure 6. Similar colour pictures have been used by Mullin and Price [Mull89] to present experimental results.

Under certain circumstances, the skeleton structure of a phase portrait defines a complete closed surface. This happens when two incommensurate frequencies are present in the original time series so that the phase portrait takes the form of a torus. Thus, the trajectory lies on a two-dimensional manifold. It is possible to build an approximation to this manifold by interpolating between adjacent segments of trajectory, much in the same way that the trajectory is constructed by interpolating between successive points. The trajectory is effectively half a wireframe. Since the array of points is ordered in time, each point in the trajectory is connected to its predecessor and successor. But there is no such simple relation for determining the other half of the wireframe.

In a region local to one point in the phase portrait, the trajectory appears as a series of approximately parallel line segments. In order to define a cross-link for this point, it is necessary to locate its nearest neighbour, subject to the restriction that the neighbour must be on a trajectory segment adjacent to but not the same as that containing the point. Another restriction must be imposed to prevent a point P from linking to point Q if Q has previously been linked to a point on the same trajectory segment as P. This eliminates the danger of mutual cross-links, where adjacent trajectory segments are linked to each other but not to their outside neighbours. This amounts to establishing a mapping from the set of trajectory points to itself, so that every point is associated with a cross-link neighbour.

If there are N points in the trajectory, then for each point of the trajectory this cross-linking algorithm involves an $O(N)$ search, giving an $O(N^2)$ efficiency overall. However, it is possible to improve the method to $O(N)$ efficiency by making use of the correlation between cross-links of successive points on the trajectory. A global cross-link search is required in two situations only, to find the neighbours $L(A)$ and $L(Z)$ of the initial and final points A and Z of the trajectory. These searches should be carried out at the beginning of the procedure. Figure 7a shows the neighbourhood of the start of the trajectory, where point A has been linked to its nearest neighbour $L(A)$. Now to find $L(B)$, the neighbour of B, only a handful of points succeeding (and including) $L(A)$ need be considered, by continuity. This inductive process can be repeated until the situation shown in Figure 7b occurs. Here, point P is cross-linked to point $L(P)$, but $L(P)$ is the last point, Z, on the trajectory. So the neighbour of the next point Q, $L(Q)$, will be totally unrelated in position to $L(P)$. However, the search for $L(Q)$ can be a local one, bounded below by the previously determined $L(Z)$. Then, in Figure 7c the beginning of the trajectory interleaves itself, and the cross-link mapping must jump to this nearer segment (beginning at point A) as soon as the mapped point reaches $L(A)$. The establishment of the table of cross-links is complete once the end of the trajectory is encountered, where Z is mapped to $L(Z)$, as seen in Figure 7d. An example of a surface generated by this cross-linking procedure is shown in Plate 37, which shows a cutaway section of the data set of Figure 6.

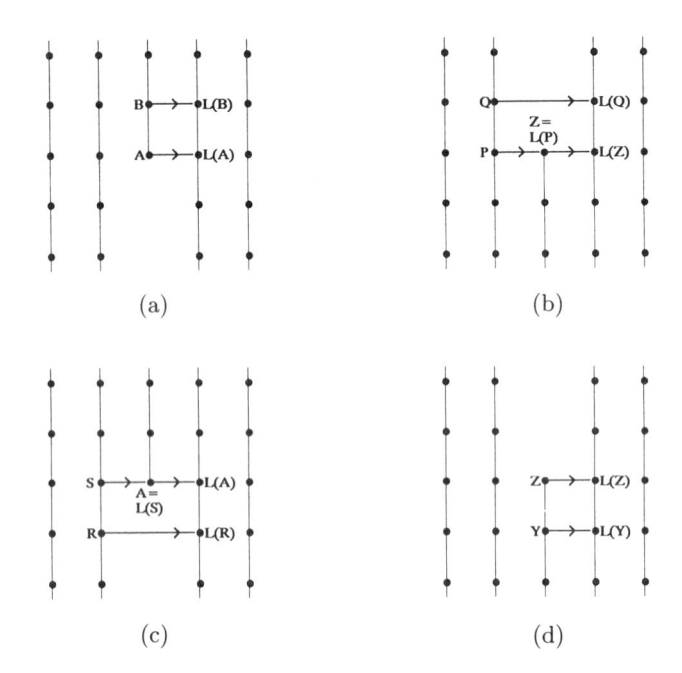

Figure 7. Defining a surface by the trajectory cross-linking procedure.

Plate 38 shows a phase portrait (from a five-dimensional numerical model of a double pendulum system) that appears to intersect itself in three-dimensional space. This indicates that we are not considering a sufficient number of dimensions of phase space, so the next step is to investigate methods of presenting a four-dimensional structure. This is not an easy task, firstly because the information compression is greater in projecting from four rather than three dimensions onto a plane, and secondly because the viewer is not used to interpreting four-dimensional objects. However, given that the illusion of three-dimensional structure can be created either by rotating a purely two-dimensional image or by applying rendering techniques to a still picture, it is likely that interactive real-time, four-dimensional rotation of a shaded image would allow the viewer to understand the essential features of the four-dimensional structure of the phase portrait. The computational power required is formidable but is within reach of top-of-the-range dedicated graphics workstations. We intend to implement and compare methods of four-dimensional image presentation in the near future.

Acknowledgements. This work was carried out in collaboration with Drs. R. Jones and D.S. Broomhead at Royal Signals and Radar Establishment, Malvern, and was supported by the SERC, Nonlinear Initiative. One of the authors (TJP) is supported by a CASE award sponsored by the Meteorological Office.

REFERENCES

[Broo86]
Broomhead, D.S., and King, G.P., Extracting qualitative dynamics from experimental data, *Physica*, Vol. 20D, p. 217, 1986.

[Casd89]
Casdagli, M., Nonlinear prediction of chaotic time series, *Physica*, Vol. 35D, p. 335, 1989.

[Fole82]
Foley, J.D., and VanDam, A., *Fundamentals of Interactive Computer Graphics*, Reading, MA: Addison-Wesley, 1982.

[Golu83]
Golub, G.H., and VanLoan, C.F., *Matrix Computations*, Baltimore: Johns Hopkins Univ. Press, 1983.

[Lore63]
Lorenz, E.N., Deterministic non-periodic flow, *Jour. Atmospheric Sci.*, Vol. 20, p. 34, 1963.

[Mull89]
Mullin, T., and Price, T.J., An experimental observation of chaos arising from the interaction of steady and time dependent flows, *Nature*, Vol. 340, p. 294, 1989.

[Mull90]
Mullin, T., Finite dimensional dynamics in Taylor-Couette flow, *IMA Jour.*, accepted for publication, Vol. 40, 1990.

[Pack80]
Packard, N.H., Crutchfield, J.P., Farmer, J.D., and Shaw, R.S., Geometry from a time series, *Phys. Rev. Lett.*, Vol. 45, p. 712, 1980.

[Ross76]
Rossler, O.E., An equation for continuous chaos, *Phys. Lett.*, Vol. 57A, p. 397, 1976.

[Schu84]
Schuster, H.G., *Deterministic Chaos*, Weinheim: Springer-Verlag, 1984.

[Take81]
Takens, F., No. 366 in Lecture Notes in Mathematics, Berlin: Springer-Verlag, 1981.

Data Visualisation Techniques for Nonlinear Systems

David Pottinger

Abstract

Data visualisation techniques, such as 3D computer graphics and animation, can play a useful role in obtaining an improved understanding of the geometric aspects of nonlinear systems. Two different physics problems, excited pendula and the 3D scattering of magnetic monopoles, are used to demonstrate this approach.

Introduction

Over the past fifteen years there have been significant new advances in the analysis of physical systems modelled by sets of nonlinear (differential) equations [Guck83; Thom88; Lamb80; Dodd82]. The introduction of geometric ideas, in a very general sense, has played a key role in this development. Consequently, it is natural to expect that data visualisation techniques, such as 3D computer graphics and animation [Burg89], can provide powerful tools to aid in the advancement of this process.

This paper briefly describes two research projects that I have been involved in at the IBM United Kingdom Scientific Centre (UKSC) that have attempted to explore and develop this relationship. The overall aim was to develop a set of general purpose data visualisation tools for nonlinear systems that would be of use, with suitable amendments, to a range of practicing scientists and engineers. By experimenting with different ways of visualising scientific data in model problems, one can hope to develop new or improved methods. This experience can then be directly applied to problems which are numerically intensive, and on which one would not normally want to spend large amounts of time with graphics experiments.

Graphics at the IBM UK Scientific Centre

I start with a few words about the graphics facilities used in these projects. The primary tool was an experimental 3D computer graphics program known as WINSOM (WINchester SOlid Modeller) [Burr89]. The aim was to provide a relatively easy-to-use 3D graphics facility that could (a) deal with a broad range of scientific data and (b) was robust enough to cope with large amounts of data.

WINSOM provides an example of a graphics system built using *constructive solid geometry* (CSG) [Burg89; Mant88]. In this approach to 3D graphics, objects are defined as sets of points, and the standard set-theoretic operations of union, difference and intersection can be used to define new, more complex objects. WINSOM has a large set of basic shapes (*primitives*) that can be 'called', e.g., sphere, cylinder, etc. A selection of some of the more commonly used primitives is given in Plate 39. In WINSOM, if one wanted to draw a piecewise-constant continuous curve from an ordered set of points, one would write something like

```
let curve =   (cylinder(r, point2) at point1) union
              (cylinder(r, point3) at point2) union
              ...
              (cylinder(r, pointn) at point(n+1));
draw curve;
```

where r is the radius of the solid cylinder; the rest is hopefully self-explanatory. If a smoother curve is required, one can insert a sphere of radius r between every pair of cylinders. Clearly, extremely complex objects can be produced in this way. A variant on this simple approach was used extensively in Example 1 in the fourth section of this paper, to produce pictures known as phase portraits. The typical number of primitives used to construct these images was rather large (in the region of 20,000 to 30,000).

WINSOM can also deal with *fields* (quantities that depend on space and time, and often occur in problems in physics, chemistry and engineering). In particular, there is the ability to draw (render) three-dimensional field contours, i.e., surfaces over which the value of some physical quantity is constant [Quar87]. If the shape of these contours changes as a parameter in the theory is varied, then a visualisation of this process can yield valuable insights into the underlying field rearrangements, which in turn may have an illuminating geometric significance. This property was used extensively in Example 2 later in the paper to investigate the shape of the energy field surrounding two magnetic charges (objects that occur in high-energy physics).

Within WINSOM, there are also facilities to account for highlighting and shading, multiple light sources, surface qualities (such as gloss) as well as perspective. Such features are necessary if one wishes to generate images with a convincing 3D 'feel'.

Clearly, if one needs to deal with large amounts of information then it is important to have good data handling facilities, e.g., those offered by a relational database. In addition, in order to quickly scan the large number of viewing possibilities available, as well as the different choices of variables for coordinate axes in multi-dimensional problems, one also needs the possibility of generating *interactive* wireframe pictures. These facilities are also available at the IBM

UKSC. Finally, there is the important possibility of animating WINSOM images to make videos.

In the above, for obvious reasons, I have restricted my attention to the various hard- and software graphics facilities available at the IBM Scientific Centre. It goes without saying that there are a variety of other graphics systems available on the market that can perform similar functions, albeit with perhaps different techniques. In addition, graphics is changing at a rapid rate through the introduction of special purpose hardware, the incorporation of parallelisation and other developments.

Scientific Data Visualisation

The practising scientist or engineer is often in the situation where he must interpret large amounts of data, obtained either from gathering experimental results or else by way of numerical computations/simulations arising from mathematical modelling. This data has much direct as well as potential information stored in it. The question then arises as to what are useful ways of extracting this information. One obvious method that springs to mind is to somehow *visualise* it. The precise way one can do this depends on the imagination of the researcher, as well as on availability of the necessary hard- and software. At the lowest level, data visualisation is simply a very convenient way of presenting data in an easily assimilable form. At its highest level it suggests unexpected connections or motivates new lines of enquiry. The minimum requirement, however, is that it be genuinely useful. Whether or not the pictures obtained are pleasant to look at is, from a scientific viewpoint, a separate aesthetic consideration (except that it might get someone interested in the problem!).

The extension from 2D to 3D graphics is quite significant, since it allows a whole new range of possibilities to be explored. It is also important to remember that most problems deal with *multi-dimensional data*. Consequently, it is very important to devise novel extensions of basic 3D graphics to allow for the incorporation of the information stored in the 'hidden' dimensions. Ways of doing this include using texture mappings as well as the more familiar colour. If this is not done, many important connections may remain undiscovered. The range of possibilities in this area have not yet been fully explored. However, by investigating a number of different physical problems, useful techniques can be expected to emerge that will have a reasonably broad range of applicability.

Examples

In this section, the foregoing remarks are illustrated by giving two examples of data visualisation, each requiring different methods. In both cases, I think it fair to say that nontrivial insights into rather difficult problems were obtained by using animation and computer graphics techniques, thus proving their worth.

Pendula in Motion (or, Visualising Higher Dimensions)

It is now known that nonlinear dynamical systems can display a variety of complicated effects, including instabilities, bifurcations and chaos [Guck83; Thom88]. The motivation for our joint project was the development of new methods for the visualisation of some of these effects, in the context of a system that would be familiar to everyone.

The problem chosen involved the nonlinear motions of an excited system of two coupled pendula. One pendulum hangs down and is free to swing from side to side. The other pendulum is attached to the bottom of the first and is constrained to swing at right angles to it. The pivot point of the top pendulum is moved up and down in a regular way by a motor at a certain frequency, which can be varied. Plate 40 shows a typical configuration taken from a computer-generated video [Pott89]. The orientation of the system is specified by the angles that the pendula make with the vertical direction.

The aim of the project was to use 3D graphics techniques to gain an improved understanding of how the motions of the pendula depend on the frequency of the excitation. To this end, it is convenient to investigate what is known as the *phase portrait* of the system [Guck83; Thom88]. This is constructed by plotting the variables that describe the positions of the pendula against their corresponding velocities, as the motion changes in time, for a given value of the frequency. In the case of the coupled pendula, five variables are needed to describe the system— the two angles, their velocities, and a variable to describe the excitation. Thus, the system lives in a 'space' which has five dimensions. To avoid confusion, note that a phase portrait is *not* obtained by plotting these variables *against* time.

Starting from a given set-up of the pendula (*the initial conditions*), the system, as it evolves in time, traces out a curve in the 5D space. By making a mathematical model for the system and solving the appropriate equations on a computer, successive points on this curve can be found. The precise shape of the curve depends on the value of the frequency of the excitation, as well as the initial conditions. The details are rather involved; the interested reader is referred to papers by Mullin et al. [Mull89; Mull90] for further information and references.

For a range of frequencies, the curve (*trajectory*) generates an overall shape that appears to be rather smooth, i.e., it has a reasonably well defined surface. An interesting question arises: How does the general shape of this surface change as the frequency of the excitation is varied? Also, What is the correct physical and mathematical interpretation of such a change?

This is where the powerful visualisation properties of 3D computer graphics are of considerable help. Clearly, it is not possible to visualise the full five-dimensional space directly. However, three of the pendula variables can be used to obtain a 'reduced' phase portrait by plotting out the points the trajectory traces out in this subset of variables. The user must decide which is the most illuminating choice of variables, based on his or her knowledge and understanding of the problem.

If this is carried out using simple wireframe methods, one obtains a rather confusing picture, since the trajectory overlaps itself many times. A better scheme is, for example, to use the CSG technique alluded to earlier in this paper. Each point becomes the centre of a sphere, and successive spheres are *unioned* with a cylinder to give a smooth 3D solid segment. This process is repeated, point by point, until the end of the numerically determined trajectory is reached. The radius is given a value such that there are no holes or gaps in the resulting solid object. This simulates, in a plausible way, the effect of including the extra data one obtains if one computes even more points on the trajectory. The number of WINSOM primitives involved in a typical image created in this way was about 20,000. However, images containing as many as 100,000 primitives were also created to investigate the phase portrait of the system for an especially important value of the excitation frequency—that associated with the bifurcation point (see below).

It is clearly of interest to try to incorporate some information about the corresponding behaviour of the system in the other two dimensions. This is achieved to a certain degree by incorporating colour. One way of doing this is to give a point on the surface a colour whose 'value' is related to the corresponding value of the chosen 'hidden' variable. Obviously, this use of colour is motivated by scientific rather than aesthetic considerations, and its practical usefulness should be interpreted in that light.

Using these methods, the 'reduced' phase portrait of the system was visualised for various values of the excitation frequency. As might be expected, the shapes obtained depended on the choice of variables. We have studied in detail the case where two angles and one angular velocity are the basic three coordinates. In this case, one finds that for the starting value of the frequency one has two smooth toroids that appear to intersect one another and that represent two different *attractors* [Guck83; Thom88]. This corresponds to the fact that there are two solutions to the equations for this value of the frequency. The intersection is an artifact of the projection from five (in which they do *not* intersect) to three dimensions. As the frequency is reduced further, one finds that the two attractors come together in a special way and join. This special splitting/joining effect is known as a *gluing bifurcation*: gluing because the two attractors get stuck together, and bifurcation because the number of solutions has changed. The new attractor is known as a double torus attractor—the trajectory first winds around one torus and then around the other. Plates 41 to 43 show these effects for the choice of the coordinates—top angle, bottom angle and its velocity (coordinate choice *A*). The surfaces have been coloured according to the *sign* of the velocity of the top angle, in order to incorporate information from one of the 'hidden' dimensions. The two tori (red/yellow and blue/green) come together and merge into one (blue/green) as the frequency is reduced. Plate 44 corresponds to Plate 42 but with the surface coloured according to the *magnitude* of the velocity of the top angle. Plates 45 and 46, corresponding to Plates 42 and 43, show the bifurcation in terms of the alternative variables—bottom angle, top angle and its velocity (coordinate choice *B*). The surfaces have been coloured according to the sign of the velocity of the bottom angle.

Clearly, 3D computer graphics allows a vivid visualisation of the bifurcation. It has also motivated an improved understanding of this process. For instance, the interpretation of the 'lip' around the edge of the two separate attractors as they approach each other has been clarified [Mull90].

Finally, a computer-generated video of this interesting bifurcation effect has been made [Pott89] (and is available on request). It shows in detail how (by an animation in time) a phase portrait is generated. It also shows (by an animation in frequency) how attractors come together and join.

Currently, together with Peter Quarendon and Iain Rodrigues, I am investigating ways of introducing *two* variables onto a 3D surface, using texture mappings. This allows, at least in principle, an additional dimension to be included and so can potentially lead to some new and interesting insights. This possibility was originally motivated by some computer art work being carried out at the UKSC [Lath89]. An experimental example of this type of work is given in Plate 47.

MONOPOLES IN MOTION (OR, VISUALISING CHANGES IN GEOMETRY/TOPOLOGY)

In the pendulum example above, the state of the system at any given time is completely specified by giving the values of just five variables (*degrees of freedom*). However, for a very large class of physical systems one is forced to deal with situations which are, of necessity, described by an infinite number of degrees of freedom. Typically this occurs when one deals with *fields*, for example the familiar electric and magnetic fields. Clearly, different types of visualisation techniques are required for this class of problem. Since the current example is, from both a mathematical and a physical point of view, extremely complicated, I only sketch the main points here. Further information and references are found in the paper by Atiyah et al. [Atiy89].

First I discuss the physics. The concept of the magnetic monopole goes back to the origin of the science of magnetism. Although all known magnets had both north and south poles, it seemed impossible to separate them. This is to be contrasted with the behaviour of electric charges, which are relatively easy to separate. This basic experimental fact was incorporated into the unified theory of electricity and magnetism (*electromagnetism*) developed by Maxwell in the nineteenth century. However, since Maxwell's time additional forces of nature have been discovered (*the weak interactions*). These interactions are important at the nuclear level. One of the great success stories of theoretical physics in the last twenty years has been the unification of these new forces with electromagnetism [Aitc82]. This combined *electroweak* theory is, in turn, a member of a yet more general class of theories, known as Grand Unified Theories (GUTs). Such exotic theories are currently of great interest, since they are thought to be important for gaining a better understanding of the evolution of the early universe. What is surprising, however, is that a generic member of this class of models predicts the existence of magnetic monopoles.

Magnetic monopoles are objects that have elementary magnetic charge, i.e., they constitute a single north or south pole. Experiments have been set up to detect their existence, although so far with no clear success. (It may surprise you to learn that IBM has been involved in the design of the superconducting detectors for one of these projects.)

Obviously it is of great interest to understand the properties of these unusual 'particles', e.g., What is their mass, and how do they interact with each other? In order to do this, one has to be able to solve some rather complicated sets of nonlinear partial differential equations (PDEs). Normally one could only hope to do this by explicit and lengthy calculations on a supercomputer. However, because there is a strong geometric element in the construction of these theories, it has proven possible, by using advanced and state-of-the-art mathematical methods, to find *exact* static solutions to this problem (in a certain limit). This result is, from a mathematical point of view, quite remarkable and, as such, constitutes a tremendous achievement. However, the solutions are so complicated that it is quite difficult to visualise their geometric properties. This is undesirable, since the topological characteristics of the solutions are expected to give helpful insights into the subtle way that geometry and physics are interwoven in this class of theories.

Consider the case of two static monopoles at a certain separation. Starting from the exact solutions, one can compute the value of the energy (mass) of the system at each point on an appropriately chosen three-dimensional grid. With the energy in this form, one can then use the contour facility of WINSOM to generate three-dimensional closed (but not necessarily connected) surfaces over which the energy is constant. The total amount of energy within each contour can also be evaluated. Based on physical arguments, surfaces that contained 5, 10, 20 and 40 percent of the total energy of the system were studied in detail. A natural question was: How did the shape of these static 3D surfaces depend on the monopole separation?

From experience with simpler, linear theories, one might expect that at large separations, where the interaction of the fields is weakest, the typical energy surface looks like two disconnected, slightly distorted spheres representing two separated monopoles. Whilst at zero separation, one might expect a simple sphere, representing one monopole of charge two. This is what one finds with the energy surfaces surrounding a pair of electric charges, for example.

At large separations one does see two slightly distorted single monopoles. However, due to the strong nonlinear interactions, at zero separation large distortions take place that give results quite different from those that a linear theory predicts. Indeed, one finds that the energy surfaces are *toroidal* rather than spherical. Plates 48 to 50 show how the topology of the four energy surfaces changes as the monopole separation is reduced to zero. They were obtained by using the contour facility of WINSOM. A section has been removed from three of the four equipotential surfaces to reveal the inner structure. The problem of the apparent breakdown of rotational symmetry is resolved through the form of the boundary conditions of the fields at infinity.

It is also very interesting to investigate how two monopoles behave when they approach each other at some speed. Unfortunately, *exact* time-dependent solutions to the two monopole system are not known. However, a useful approximation (*the adiabatic approximation*) exists that makes use of the exact static solutions [Atiy88a]. It is valid provided that the monopoles are moving relatively slowly. Using this approximation, we have carried out a detailed numerical analysis of the scattering of two monopoles and have recorded the results in the form of a computer-generated video (which is available on request) [Atiy88b]. Clearly, animation is the most appropriate technique for this type of problem. Each of the separate frames of the movie is a separate WINSOM image.

Perhaps one of the most interesting features of the movie is the visualisation of the head-on scattering of two monopoles. One finds that they go off *at right angles* to their initial line of approach. (Just imagine playing billiards with this type of collision rule!) This unusual result can be traced back to the axial symmetry present at the centre point of collision, as illustrated in Plate 50. It turns out that the dynamic interactions of slowly moving monopoles is a rather complicated affair. Thus, the computer-generated video provides a very useful means of developing an improved understanding of this complex system. The interested reader is referred to papers by Atiyah et al. [Atiy89; Atiy88b] for further details.

It is important to realise that there is another, perhaps more fundamental, reason why this exotic sounding physics problem is of great interest. There are strong indications that the mathematical methods used to solve the monopole problem can also be used to solve other complicated 3D PDEs. This is a rather important point since, although an enormous range of physical systems are modelled by nonlinear partial differential equations, little is known about them from a theoretical point of view. When numerical computations of PDEs are done, as they often are in engineering applications for example, numerous additional approximations have to be made, simply in order to make progress. Clearly, this situation could be improved if our basic understanding of these systems improved. Consequently, all theoretical advances in PDEs are extremely important and should be investigated in great detail in order to realise their full potential. Computer graphics can play a useful role in this process.

It is interesting to speculate that there is a parallel here with the development of the theory of one-dimensional *solitary waves* (or *solitons*) [Lamb80; Dodd82]. These are solutions to certain nonlinear differential equations in one space and one time dimension. The solutions are localised in space and have the remarkable property of not undergoing a substantial change in form under interaction, i.e., they are 'waves' that behave like 'particles'. Significant progress has been made in the last twenty years in developing techniques for the analysis of 1D solitons. The range of their practical application is now quite wide and includes superconductivity, fibre optics, lasers and polymer systems [Dodd82]. Many of these advances were motivated by the results of computer experiments, investigated with the use of graphics and animation techniques [Zabu81]. In this context it is interesting to note that the magnetic monopole previously discussed may be thought of as a form of 3D soliton.

Conclusions

A wide range of physical systems can be modelled by sets of nonlinear differential equations. Unfortunately, many features of these systems are still poorly understood. Data visualisation techniques, such as 3D graphics and animation, provide useful ways of bringing the geometric features of these systems 'to life'. This often leads to improved understanding of the problem under consideration.

This paper has presented examples of the practical application of these methods to two problems in physics. The general results obtained from these projects can be summarised as follows. The type of visualisation methods that are most appropriate depend on the nature of the problem. For systems that are characterised by relatively few degrees of freedom, one can use projections to explore the multi-dimensional data space. Such an approach can be improved by appropriate incorporation of colour and texture. For systems with a large number of degrees of freedom (such as fields) alternative techniques are usually more helpful, e.g., 3D contour surfaces. Finally, in the case where a physical process depends on a continuous parameter (which may or may not be time), animation clearly provides an invaluable tool. These are, of course, just preliminary steps. Much more work needs to be done before the potential of 3D computer graphics is fully realised. However, due to the increasing use of numerically intensive computing, the relevance and practicality of data visualisation techniques can be expected to grow rapidly in the near future.

Acknowledgements

My collaborators in the first project, Pendula in Motion, were Dr. Tom Mullin and Anne Skeldon of the Nonlinear Systems Group at the Clarendon Laboratory, Oxford University, together with Iain Rodrigues and Stephen Todd from the UKSC, Winchester.

My collaborators in the second project, Monopoles in Motion, were Professor Sir Michael Atiyah and Dr. Nigel Hitchin from the Mathematical Institute, Oxford University, and Dr. John Merlin from the Physics Department of Southampton University, together with Bill Ricketts of the UKSC, Winchester.

REFERENCES

[Aitc82]
    Aitchison, I.J.R.,and Hey, A.J.G., *Gauge Theories in Particle Physics*, Bristol, UK: Adam Hilger, 1982.

[Atiy88a]
    Atiyah, M.F., and Hitchin, N.J., *The Geometry and Dynamics of Magnetic Monopoles*, Princeton, NJ; Princeton Univ. Press, 1988.

[Atiy88b]
    Atiyah, M.F., Hitchin, N.J., Merlin, J.H., Pottinger, D.E.L., and Ricketts, M.W., Monopoles in motion, IBM UKSC video, 15min., 1988.

[Atiy89]
Atiyah, M.F., Hitchin, N.J., Merlin, J.H., Pottinger, D.E.L., and Ricketts, M.W., Monopoles in motion: a study of the low-energy scattering of magnetic monopoles, IBM UKSC Report 207, 1989.

[Burg89]
Burger, P., and Gillies, D., *Interactive Computer Graphics*, Reading, MA: Addison-Wesley, 1989.

[Burr89]
Burridge, J.M. et al., The WINSOM solid modeller and its application to data visualisation, *IBM Sys. Jour.*, Vol. 28, pp. 548–568, 1989.

[Dodd82]
Dodd, R.K., Eilbeck, J.C., Gibbons, J.D., and Morris, H.C., *Solitons and Nonlinear Wave Equations*, London: Academic Press, 1982.

[Guck83]
Guckenheimer, J., and Holmes, P., *Nonlinear Oscillations, Dynamical Systems, and Bifurcations of Vector Fields*, New York: Springer-Verlag, 1983.

[Lamb80]
Lamb, G.L., Jr., *Elements of Soliton Theory*, New York: Wiley Interscience, 1980.

[Lath89]
Latham, W.H., and Todd, S.J., Technical note—Computer sculpture, *IBM Sys. Jour.*, Vol. 28, pp. 682–688, 1989.

[Mant88]
Mantyla, M., *An Introduction to Solid Modelling*, Rockville: Computer Science Press, 1988.

[Mull89]
Mullin, T., Pottinger, D.E.L., and Skeldon, A., The five-dimensional pendulum picture show, *New Scient.*, No. 1689, pp. 46–47 and front cover, 1989.

[Mull90]
Mullin, T., Pottinger, D.E.L., Skeldon, A., Rodrigues, I., and Todd, S., Phase portraits for parametrically excited pendula: an exercise in multidimensional data visualisation, IBM UKSC Report 213, 1990.

[Pott89]
Pottinger, D.E.L., Rodrigues, I., and Quarendon, P., Pendula in motion, IBM UKSC video, 8min., 1989.

[Quar87]
Quarendon, P., A system for displaying three-dimensional fields, IBM UKSC Report 171, 1987.

[Thom88]
Thompson, J.M.T., and Stewart, H.B., *Nonlinear Dynamics and Chaos*, New York: John Wiley and Sons, 1988.

[Zabu81]
Zabusky, N.J., Computational synergetics and mathematical innovation, *Jour. Comp. Phys.*, Vol. 43, pp. 195–249, 1981.

Index

Active memory, 132
Addressing methods, image representation and, 135
Adiabatic approximation, for magnetic monopole system, 266
Affine mapping, 134–138, 146
Aliasing effects, 114
Animation
  of clouds, 116
  coupled pendular bifurcation effect, 264
  of magnetic monopole system, 266
  for Winchester solid modeller, 261
Artistic creativity, chaos theory applications, 211, 216–222
Arts and Crafts Movement, 221
Associative learning automata, 157–163
Attraction rule, for Barnsley chaos game, 16
Attractor(s)
  chaotic pendulum motion, 199
  coupled pendula system, 263
  double torus, 263
  of IFS, 123–126, 146
  Lorenz system, 200–202
  of meta-IFS, 128
  Rossler attractor, 250
  tile structure representation and, 123–126
Attractor basin, 37–38
Axis of symmetry, for Mandelbrot sets, 39–40
Axonometric orthographic projection, 253

Bach, J. S., 8
Barnsley, Michael, 15
Barnsley chaos game, 15–18
Bifurcation point, 3D data visualization system, 263

Binary numbers, 11
Biological objects, fractal growth model, 71, 76–86
Borel subset, 155
Boundary constraints, for Dielectric Breakdown Model, 55
Brownian motion
  approximation by spatial methods, 96–100
  approximation by spectral synthesis, 101–102
  fractional, 96–110, 120
  multidimensional extension, 104
  one-dimensional, 91–106
  random fractals and, 91–102
Butterfly effect, 196, 198

Canonical Mandelbrot set, 36
Cantor, G., 9–11
Cantor comb, 9–11
Cantor set, 11, 14
  dimensionality of, 27–28
  IFSs and, 146, 156
Cardiff, growth model for, 46, 60, 63–64
Cayley, Arthur, 20
Cellular automata, 72, 219
Center of curvature, phase portrait representation and, 254–255
Central business districts, 48
Central place theory, 44
Chaos
  basic features of, 195
  sensitivity to initial conditions and, 196–198
  transient, 231;
  See also specific problems
Chaos game, 15–18
Charged particle interactions, 225, 229–235, 260, 264–266
City structures, morphology of, 43–66

Climatology, strange attractors and, 202
Clouds, random fractal simulation, 116
Clusters, urban growth models and, 49, 52
    constraints on space for, 61–63
    Dielectric Breakdown Model and, 55
    DLA model, 53–58
    edge effects, 59
    estimating fractal dimension of, 58–61
    fractal signature of, 60–61, 64
    multifractal nature of, 60
    potential theory and, 55
    scale of, fractal dimension and, 65, 66
    urban growth and, 49, 52
Coastline problem, 43
Collage theorem
    artistic applications of, 219
    tiling and, 124
Color
    map, pixel plane definition, 131
    random IFSs and, 149–150
    3D data visualization system, 263
Complex numbers, 20–22
Complex plane, 19
    Julia sets and, 22, 37
    Mandelbrot set and, 26
Complex quadratic mapping, 72
Computer art, 264
    chaos theory applications, 211, 216–222
    IFS applications, 142
Computer-assisted design (CAD), IFSs and, 138, 142
Computer-generated video, 264, 266
Computer graphics
    biological object visualization and, 71
    Distributed Array of Processors, 131–132
    fractal tiling structures in, 121–122
    Iterated Function Systems, 136–138, 146–147, 150
    Julia set representations, 22–24
    nonlinear systems analysis and, 193, 259–267
    phase portrait representation, 252–256

Computer graphics (Cont.):
    random IFS and, 150
    tiling structure representation, 130
    urban morphological applications, 45
Condensation set, 149, 152
Constructive solid geometry (CSG), 260, 263
Continuous variables
    fractal resolution and, 100
    random IFSs and, 155
Contraction mapping, 122, 128
Contractivity, 122, 125, 128
Convective overturning, 200
Coral species, growth forms of, 77
Creativity, chaos theory applications, 211, 216–222
Cross-linking algorithm, 255
Curvature, center of, phase portrait representation and, 254–255
Curve fitting, phase portrait display and, 253
Cyclotron frequency, 231–232

Dancing dot method, for generating IFS, 128–129
Database, relational, 260
Data compression, 145
Data visualization, for nonlinear systems, 259–267; See also Computer graphics; specific methods, problems
Dedekind, J. W. R., 11
Degrees of freedom,
    system state characterization and, 264
Delays, method of,
    for phase portrait construction, 249, 250
Dendritic forms
    cluster growth constraints, 62
    Dielectric Breakdown Model and, 56
    DLA model and, 53
    urban development and, 48
Density, for fractal urban morphology model, 52, 54, 58, 59, 61, 62, 66
Depth cueing, 254
Design, chaos theory applications, 211, 216–222

Deterministic systems
  in DLA model, 53
  Iterated Function Systems and,
    126–127, 146–149
  processing speed and, 136
  tiling structures and, 119, 120,
    126–127
  of urban growth, 46, 48
Dielectric Breakdown Model (DBM),
  55–58, 63–64
    constraints on space for, 61–62
    estimate of fractal dimension for, 59
Differential equations, 265, 266
  linear systems and, 194
  Lorenz set, 200, 211–212
  nonlinear, 200
Diffusion Limited Aggregation (DLA),
  46, 53–64, 72
Digital image generation, Distributed
  Array of Processors, 131–132
Dimension, 26–27
  extension of, for Fourier filtering
    method, 107–110
  of fractal objects, 27–28
  fractional, 44;
    See also Fractal dimension
  generalizing displacement methods
    for, 103–106
  Hausdorff's definition of, 28–30
  of phase space, 225
  scientific data visualization, 261
Discretization, for calculation of
  Mandelbrot set, 36
Distance diminishing model, learning
  automata and, 155
Distributed Array of Processors (DAP),
  131–132
DLA, See Diffusion Limited Aggregation
Double torus attractor, 263
Dragon curve, 13
Dürer, Albrecht, 8, 16, 18
Dynamical systems
  attractors in, 200–202
  charged particle systems, 225,
    229–235
  nonlinear differential equations and,
    200
  phase space analysis, 199, 247–249
  3D visualization, of coupled
    pendula, 262–264

Edge effects, for fractal growth
  model, 59
Electronic oscillator, 239–240
Electroweak theory, 264
Equipotential surfaces, 265
Escher, M., 8
Euclid, 18
Euler, Leonard, 194
Euler method, 230

Fatou, Pierre, 19, 22, 24
Field contours, for WINSOM 3D
  modeller, 260
Field of influence, for DLA model, 53
Filled-in Julia sets, 22, 23
Finite length effects, Taylor-Couette
  flow and, 242
Five-dimensional space, coupled
  pendular motion in, 262
Fluid dynamics
  attractors in, 202
  nonlinear systems, 194–195
  Taylor-Couette flow, 240–245, 250
Forward difference method, 58
Fourier filtering method, 90, 102–103
  in higher dimension, 107–110
Fractal(s)
  artistic applications of, 219
  historical examples of, 8
  Lorenz attractor and, 202
  Mandelbrot's definition of, 7, 71
  natural vs. synthetic, 30
  non-physical applications, 44
Fractal dimension, 26–27, 44
  approximation of, 29
  of biological structures, 71, 76
  of Brownian graph, 96
  of clusters, 54, 58–61, 65, 66
  for DLA, 54
  using Fourier filtering method, 110
  from log-log regressions, 58–59
  rescale-and-add method, 115
  of urban structures, 49, 51, 58–61
  variable function for, 112
Fractal dust, 38
Fractal growth, See Growth processes
Fractional Brownian motion, 96–110,
  120
Functional-based modeling, 90, 110

Galton, Sir Francis, 196
Gasket, Sierpinski, 14–18, 24, 27–28
Generator function, for biological growth model, 74, 78–79
Geography, urban, fractal simulation of, 43–66
Geometric constructions
  Dürer's pentagons, 8–9, 16
  growth models and, 72–73
  Sierpinski fractal objects, 14–18, 24, 27–28
Gleick, James, 204
Gluing bifurcation, 263
Godel, K., 8
Golden section, 8
Gradient governed growth, 56
Grammar, of linguistic model, 217–218
Grand Unified Theories (GUT), 264
Graphics, See Computer graphics
Grids, See Lattices
Growth processes
  Cardiff example, 63–64
  constraints on space for, 61–63
  Dielectric Breakdown Model, 55–58
  Diffusion Limited Aggregation (DLA) model, 46, 53–58
  fractal dimension, 76
  irreversible, 46, 52
  ramifying fractals and, 73–75
  sponge, fractal model of, 71, 76–86
  urban morphology and, 43–66

Hamiltonian systems, 225–235
Hausdorff, Felix, 28–29
Hausdorff dimension, 7, 27;
  See also Fractal dimension
Hausdorff metric, 123, 146
Hermite interpolation, 112
Hidden-surface removal, phase portrait representation and, 254
Hierarchical IFSs, 128
Hierarchical ordering, of city distributions, 44
Hierarchical tiling structures, 126
Highlighting, for Winchester solid modeller, 260
Hilbert, D., 11
Hofstadter, D., 8
Hutchinson metric, 150

Hydrodynamics, See Fluid dynamics
Hyperbolic Iterated Function System, 146

Image synthesis
  addressing methods, 135
  interactive system for (ISIS), 138–142
  random fractals in, 89–116
  tiling structures, IFSs and, 119–142;
  See also Computer graphics
Imaginary number, 20
Initial conditions, system sensitivity to, 196–198
Initial values plane, 232–233, 235
Initiator, for morphological growth model, 73
Interactive System for Image Synthesis (ISIS), 119, 120, 138–141
Interpolation
  phase portrait representation and, 253
  of points in midpoint displacement, 99
  random fractal point-evaluation, 112
Irreversibility, of urban growth, 46, 42, 63, 64, 65
ISIS, See Interactive System for Image Synthesis
Iterated Function systems (IFSs)
  artistic applications of, 219–220
  dancing dot method for generating, 128–129
  deterministic processes, 126–127
  as graphical objects, 136–138
  hierarchical, 128
  learning automata and, 145
  neural networks and, 145, 147–149, 157–163
  procedural modification, 211
  random processes, 127–130, 149–152
  stochastic learning automata and, 145, 152–162
  tiling structure generation and, 119, 122–142
  visual aspects of, 211

Julia, Gaston, 19, 22, 24
Julia sets, 22–24
  calculation of, 37–38
  graphic images of, 219
  Mandelbrot sets and, 25–26, 40
  random fractal approximation and, 90

KAM theorem, 226–227
Koch, Helge von, 11
Koch curve, 11–13, 18
  dimensionality of, 27–28
  IFSs and, 147
  rewriting system for, 123fig.

Lacunarity, 99, 101fig., 114
Laminar flow, 240
Landscapes, midpoint method and, 105
Laplace's equation, 55, 58
Laplacian fractals, 56
Lattices
  DLA model and, 53–54, 57
  extending dimension of, 104, 105
  nonrectangular, 105
Learning automata, 145, 152–163
Length effects, for Taylor-Couette flow, 242
Levy's curve, 13
Life and death techniques, 219
Linear mapping, 73
Linguistic model of creativity, 217–218
Lipschitz constant, 122
Local indexing, 132
Local splines, 253
Log-log regression, 58–59
Lorentz force, 229
Lorenz, E. N., 200, 211
Lorenz attractor, 200–202
Lorenz set, 200, 211–212
L-systems, 72, 121
Lyapunov exponents, 253

Mackintosh, C. R., 211, 220
Magnetic field, charged particles in, 225, 229–235
Magnetic monopoles, 264–266

Mandelbrot, Benoit, 7, 22, 26, 43, 71
Mandelbrot sets, 24–26
  axis of symmetry for, 39–40
  calculation of, 35–37
  canonical, 36
  defined, 25
  graphic images of, 219
  Julia sets and, 40
Mandelbrot-Weierstrass function, 90, 112
Mapping
  affine, tile structure representation, 134–138
  contraction mapping, 122
  linear, 73
Markov chain, 128, 129
Martin, B., 213–215
Maxwell, J. C., 211
May, R. M., 203
Memory
  Distributed Array of Processors, 131–132
  nth-order randomness and, 218
Menger sponge, 14
Metaskew, 137, 141
Metatiling structures, 120, 124
Midpoint displacement methods, 90, 120
  fractional Brownian motion and, 92–94, 97–99
  generalizing to higher dimensions, 103–106
  interpolated point displacement, 99–101
Monkey tree curve, 13
Monopoles, 264–266
Morphology
  sponge growth model, 71, 76–86
  of urban systems, 43–66
Multi-dimensional data, 261
Multifractals, 55, 60

Natural phenomena
  linear systems models, 194
  random fractals and representation of, 89
  sponge growth model, 71, 76–86
  strange attractors and, 202
Navier, Claude, 194
Navier-Stokes equations, 194–195, 240

Neighborhood
  Hausdorff's concept of, 28–29
  urban structures and, 51
Neummann, John von, 195
Neural networks, Iterated Function Systems and, 145, 147–149
Noise
  associative learning automaton and, 162
  Dielectric Breakdown Model and, 56
  one-dimensional function, 112
  singular value decomposition, 252
  stochastic synthesis, 90
  white, fractional Brownian motion and, 91, 102
Nondissipative processes, particle-field interactions and, 229
Nonlinear differential equations, 200, 265, 266
Nonlinear electronic oscillator, 239–240
Nonlinear mapping, Mandelbrot set calculation and, 35–37
Nonlinear systems
  computer methods and, 193
  coupled pendula and, 262–264
  historical backgrounds of, 194–195
  magnetic monopoles, 264–266
  multiplicity of states in, 242
  sensitivity to initial conditions, 196–198
nth-order randomness, 218–219
Nyquist limit, 114

One-point correlation functions, 58
Onset of turbulence, 195, 240–245
Orthographic projection, phase portrait representation and, 253
Oscillation
  charged particle and, 229
  nonlinear electronic oscillator, 239–240
  parametrically excited pendulum and, 238–239
  phase portraits of, 199
  Taylor-Couette flow and, 241, 250
Oversampling factor, 253

Parallelism, 119, 131–132, 134, 235
Parameter space, 233–234, 239
Parametrically excited pendulum, 238
Partial differential equations, 265, 266
Particle-field interactions, 225, 229–235
Peano, G., 11
Pendulum, motion of, 195
  attractor for, 200
  chaotic behavior, 195
  coupled, 3D visualization of, 262–264
  nonlinear differential equations for, 200
  parametric excitation of, 238–239
  phase portrait for, 199, 248
  periodic perturbation and, 227
Pentagonal constructions, 8–9, 16, 18
Period doubling, 198, 237–238, 239, 243
Periodic perturbation, pendulum motion and, 227
Perlin, K., 110–112
Perron-Frobenius theorem, 152
Perspective, in Winchester solid modeller, 260
Perturbation
  charged particle system, 232–234
  KAM theorem and, 226
  pendulum motion and, 227
Phase portrait, 199
  construction of, using method of delays, 249, 250
  coupled pendula and, 262
  graphical presentation of, 252–256
  nonlinear electronic oscillator, 239
  qualitative structure of, 253
  for simple pendulum, 248
  for Taylor-Couette flow, 243, 250
  3D representation of, 253–254
  time series analysis, 249–252
Phase space, 199, 247–249
  dimensionality of, 225
  initial values plane, 232–233, 235
Philomorphs, 44
Pixel, Julia set representation and, 22–24, 26
Pixel plane, 130–131, 134
Planets, fractal approximations of, 96
Plasma physics, 229
Poincaré, Henri, 8, 27, 197, 211

Poincaré mapping, 226, 231
Point evaluation, of random fractal
   function, 110–116
Popper, K., 7
Population growth equation, 195,
   198–199
Postprocessing functions, for biological
   growth model, 75, 79
Potential theory, fractal growth and, 55,
   57, 58
Power law, spectral density of fractional
   Brownian motion, 110
Presentation, of linguistic model of
   creativity, 217–218
Primitives, for Winchester solid
   modeller, 260, 263
Probability distribution growth model
   and, 55–56
      IFS-type learning automaton, 162
Procedural modification, IFSs and, 211,
   219
Processing speed, deterministic vs.
   random processes, 136
Production rule-based techniques, 219
Pseudorandom numbers, 212
Pythagoras tree, 18–19, 21fig., 30

Quadratic map, 237–238

Radiative accretive architecture, 77
Ramifying fractal, growth models and,
   73–75
Random addition method, 105
Random cut method, 90, 95
Random fractals
   approximation of, 89–90
   Brownian motion, 91–102
   extension to higher dimensions,
      103–110
   in image synthesis, 89–116
   midpoint displacement method and,
      92–94
   point evaluation of, 110–116
   rendering aspects of, 90
   rescale-and-add method, 114–116
Random Iterated Function systems,
   149–152
Random numbers, 212

Randomness
   artistic applications, 211, 218
   learning automata and, 152
   local indexing, 132
   naturally occurring fractals and, 30
   nth-order randomness and, 218–219
   processing speed and, 136
   tiling structures and, 119, 120,
      127–130
Rank size rule, for cities, 44
Raster device, Julia set representation
   and, 22–24, 26
Rectangular grids, alternatives to, 105
Recursive equations, visual effects of,
   213–216
Reeve, D., 24
Relational database, 260
Rendering, random fractal applications,
   90
Rescale-and-add method, 90, 114–116
Resolution
   in midpoint displacement method,
      99
   of random fractals, 89–90
Reward-punishment scheme, for learning
   automata, 156
Rewriting systems, tiling structures and,
   121
Reynolds number, 241
Ribbon method, phase portrait
   representation and, 255
Richardson, L. F., 43, 195
Rossler equations, 248–250
Rotational symmetry, 40
Rule-based techniques, 219
Runge-Kutta method, 230

Sampling rate, phase space
   representations, 252–253
Scale
   clusters, fractal dimension and, 65,
      66
   coastline length problem, 43
   DLA clusters and, 54
   fractal urban growth model, 62
   fractional Brownian motion and,
      102
   metatiling structures and, 124
   urban morphological

276 Index

Scale (Cont.):
    representations and, 47–48, 49–50
Scattering, in magnetic monopole system, 266
Schizophrenia, 203
Scientific data visualization, 261
Screening length, for Dielectric Breakdown Model, 57
Seed sites, for fractal urban growth model, 51, 53, 64, 65
Selector, of linguistic model of creativity, 217–218
Self-similarity property, 7, 27, 44, 202
    Brownian motion, 91, 96
    in charged particle system, 233
    Collage technique and, 219
    DLA clusters and, 53–54
    Iterated Function Systems and, 119
    of urban structures, 48, 49–50
Self-tiling structures, 120, 122, 124, 132
Separatrix, 227
Set theory, fractal properties, 9
Shading, 254, 260
Sierpinski, Waclaw, 13
Sierpinski carpet, 14, 18
Sierpinski sponge, 14, 27–28
Sierpinski tetrahedron, 14
Sierpinski triangle (or gasket), 14–18, 24
    dimensionality of, 27–28
    IFSs and, 146
Signature, of cluster, for fractal urban growth model, 60–61, 64
Single Instruction Multiple Data (SIMD) parallelism, 119, 120
Singular value decomposition (SVD), 250–252
Skeleton structure
    of phase portrait, 255
    for sponge growth forms, 77
Social physics, 44
Solid texture, 110
Solitary waves (solitons), 266
Space filling curves, 11
Spectral density, 104, 107, 112
Spectral exponent, 101–102
Spectral synthesis, 101–106
Sponge, fractal growth model, 71, 76–86
Sponge, Sierpinski, 14, 27–28
Stability, KAM theorem and, 226
Stationarity, of random function, 99

Stochastic bands, 227, 234
Stochastic learning automata, 145, 152–162
Stochastic models, fractal urban growth simulation, 63
Stochastic noise synthesis, 90
Stokes, George Gabriel, 194
Strange attractors, 200–202
Successive random addition, 99
Symmetry, for Mandelbrot sets, 39–40

Taunton, England, 57, 60, 63
Taylor-Couette flow, 240–245, 250
Tetrahedron, Sierpinski, 14
Texture, solid, random functions and, 110
Texture mappings, 261
Three-dimensional representation
    coupled pendula and, 262–264
    nonlinear systems, UKSC research, 259–267
    of phase portraits, 253–254
Tiling structures
    computer graphics representation, 121–122, 130
    deterministic processes, 126–127
    definition of, 120
    hierarchical, 126
    Iterated Function Systems and, 119–142
    random processes, 127–130
Time series analysis, phase portrait and, 249–252
Toroidal forms, 227, 231, 234, 243, 263, 265
Trajectory, 225–226
    coupled pendula and, 262
    particle-field system, 230–232
    periodic perturbation and, 227
    phase portrait and, 247–248, 253
    population model equation and, 198–199
    for simple pendulum, 248
    singular value decomposition, 250–252
    time series analysis and, 249
Transient chaos, 231
Transportation, urban development and, 48, 49, 62, 63

Triangle, Sierpinski, 14–18, 24, 27–28
Triangular grids, 105
Triangular subdivision method, 120–122
Turbulence, 112, 195, 240–245
Turbulent flow, 240
Two-point correlation analysis, 57, 59

Ulam, Stanislaw, 203
United Kingdom Scientific Centre (UKSC), data visualization research, 259–267
Urban growth and structures, fractal simulation of, 43–66

Vector plotters, phase portrait representation and, 253–256

Video, 264, 266
Viscous fingers, 46, 55, 56
Vocabulary, of linguistic model of creativity, 217–218
von Neummann, John, 195
Voxel-based structures, 136

Weak interactions, 264
Weather forecasting, 203, 211–212
White noise, in Brownian motion representation, 91, 102
Winchester Solid Modeller (WINSOM), 260
Wireframe methods, 263

Zeeman, E. C., 203